Screen Addiction

Why You Can't Put that Phone Down

Sister Marysia Weber, RSM, DO, MA

En Route Books & Media, LLC
St. Louis, MO

En Route Books and Media, LLC
5705 Rhodes Avenue
St. Louis, MO 63109

Cover credit: TJ Burdick

Library of Congress Control Number: 2019942933

ISBN-13: 978-1-950108-08-4
ISBN-10: 1-950108-08-2

DEDICATION

This book is dedicated to all those seeking a way to handle more effectively the compulsive cravings of our social media-driven lives.

Table of Contents

PREFACE

Silicon Valley families have an interesting conundrum: The parents work at Fortune 500 companies developing digital products for users across the globe. Their children, on the other hand, attend schools in which screens are rarely, if ever, used. You would think that the tech-driven programmers would want their children to be well-versed in code and online learning like their parents, but the truth is, they don't want their children to know how to use devices, they want them to know how to use their brains.

One of the most famous of these parents who did not allow his children to use screens at home was none other than the founder of Apple, Steve Jobs. When he created the smartphone, his intent was to combine the everyday efficiency tools we needed into a singular device. For him, this meant having the ability to talk on the phone, listen to music, send text messages, check the weather and view the internet. He never wanted to create an app store because he didn't want 3rd party programmers infiltrating his efficiency – his time was precious.

When the Apple App Store launched in 2008, Jobs reluctantly permitted the greatest "Library of Distraction" to enter the digital world. What nobody guessed was just how much this invention would affect the real world. Ten years later, we are just now realizing that the tech revolution did not just bring the freedom and efficiency we thought it would – it can also enslave us.

Today, the average human being in developed countries spends just over three and a half hours on his or her smartphone. When additional hours from alternative screens are added, the total leaps to 8-10 hours depending on each person's lifestyle. The result? We now spend most of our waking lives with our eyes glued to a screen and, during those few moments when we are not scrolling, searching and posting, our minds are usually thinking about the digital world instead of the real world around us. In short, our minds have become slaves to the technology they created.

We are social beings and, as such, we are wired for relationship with one another and with God. Over the course of several millennia, we have become masters at non-verbal cues and linguistic dialogue, and this has resulted in strong, intimate relationships with our fellow man. We've flourished as a human community by disciplining our minds and turning our meditative dreams into concrete realities. We've put a man on the moon, cured a multitude of diseases, and even managed to share in the thoughts of God Himself through contemplative prayer and attentiveness to divine revelation.

In this book, *Screen Addiction: Why You Can't Put*

that Phone Down, Sr. Marysia has compiled a magnificent treatise against the threat that looms to overtake our mental abilities to imagine, create, contemplate and relate with our fellow brothers and sisters in Christ. She analyzes the current physiological and psychological research surrounding screen addiction and blends it with the timeless theological and philosophical truths of the Catholic faith in an easy-to-understand read.

Long before the threat of digital distractions, men and women could better assimilate mental thought with the natural and metaphysical realities that surrounded them. Sr. Marysia's book aims to return the minds of God's people to that state without sacrificing the efficiency that screens can provide us. It is in our best interest to read it attentively so as to protect ourselves from the machines that have hijacked our attention spans, crippled our sense of belonging and weakened our ability to contemplate the spiritual realm.

Technology is not the saving grace that we thought it would be 10 years ago. We know that now. What we have yet to discover is how these devices may best be used to attain our singular purpose in life, which is nothing short of our salvation.

TJ Burdick, author of *Detached: Put Your Phone in its Place* and founder of Signum Dei, Catholic Learning Communities, at https://www.signumdei.com/

INTRODUCTION

Have you ever had an irresistible urge to check your texts, social media posts or email updates when your cell phone rings, beeps or buzzes? Is the amount of time spent on your cell phone increasing? Do you text, access social media or open your email account while driving? Do you find yourself mindlessly checking your cell phone many times a day even when you know there is likely nothing new or important to see? What is it about our relationship with our cell phones that makes us want to check-in with it many times a day? Are we aware of how many times we even do so?

A recent survey found that Americans check their phones 47 times a day, and 50% check their phones in the middle the night. 18 to 24-year-olds check their phones 82 times a day, and 75% of them check their phones in the middle of the night.[1]

What about our youth? How much time are they spending on their devices? Although the estimates vary, the Kaiser family foundation reports that the average 8- to 10-year-old spends almost 8 hours on

---

[1] Deloitte, 2016 Global Mobile Consumer Survey: US Edition; The market-creating power of mobile, 2016.

various digital devices while teenagers spend 11 hours in front of screens.[2] This is more time than they do anything else, including sleep.

Is this increasing number of hours spent using the cell phone having any effects on our well-being? Research is telling us that the answer to this question is a resounding yes! Excessive screen time is contributing to an increase in anxiety, depression, attention deficit and lower self-esteem, to name a few.[3] Delays in language development and social skills in young children have also been noted.[4,5]

Section one offers an overview of the psychosocial effects digital media are having on children and adults. Section two considers screen addiction and online gaming and offers several suggestions and resources to address the detrimental effects excessive use of electronic media is having on some people. Section three considers Internet pornography addiction and treatment. Section four focuses on some of the most popular digital applications and websites used by

---

[2] Henry J. Kaiser Family Foundation, Generation M2: Media in the Lives of 8- to 18-Year-Olds, January 2010.

[3] Y. J. Kim, H. M. Jang, Y. Lee, D. Lee, D. J. Kim. Effects of Internet and Smartphone Addictions on Depression and Anxiety Based on Propensity Score Matching Analysis. Int J Environ Res Public Health. 2018 Apr 25;15(5).

[4] H.L. Kirkorian, K. Choi, T. A. Pempek, Toddlers' word learning from contingent and noncontingent video on touch screens. Child Dev.2016;87(2):405–413.

[5] Jenny S. Radesky, Dimitri A. Christakis, Increased Screen Time Implications for Early Childhood Development and Behavior Pediatr Clin N Am 63 (2016) 827–839.

teenagers and what parents need to know in order to provide a safety net for their children. Section five addresses the question, "Is electronic media making us smarter?" Section 6 briefly overviews new research that engages the human genius in creating apps that promote neurocognitive health. A description is provided of the largest government study underway, which will follow children into adulthood to garner a better understanding of the interplay between childhood experiences and screen use. Section 7 describes the neurobiology of joy as an essential ingredient for healthy growth and development. The appendices contain three discussion booklets—one for parents, one for young persons and one for teens with age appropriate questions, tips, suggestions and resources. The booklets may also be downloaded from https://enroutebooksandmedia.com/screenaddiction as stand-alone documents for use by various groups. There is also a link on the website and in the appendices to a 5 ½ minute video vignette on screen addiction to help focus the discussion.

Section 1

Youth and media—When does digital use begin and does it really matter?

Infants naturally gravitate toward mobile media in the home. They often observe their parents using mobile devices. The flashing of screens draws their attention. They accidently discover that touchscreens facilitate more media encounters. According to a study by the American Academy of Pediatrics, children today begin interacting with digital media at four months of age. 44% of children under age one use a mobile device on a daily basis to play games, watch videos or engage apps. This percentage increases to 77% of two-year-olds.[6]

---

[6] J. S. Radesky, M. Silverstein, B. Zuckerman, D. A. Christakis, Infant self-regulation and early childhood media exposure Pediatrics. 2014 May;133(5):e1172-8. doi: 10.1542/peds.2013-2367. Y. L. Reid Chassiakos, J. Radesky, D. Christakis, et al., AAP Council on Communications and Media. Children and Adolescents and Digital Media. Pediatrics. 2016;138(5): e20162593. H. K. Kabali, M. M. Irigoyen, R. Nunez-Davis, et al. Exposure and use of mobile media devices by young children. Pediatrics. 2015;136(6):1044–1050. G. Geng and L. Disney, "A Case Study:

Another study demonstrated that preschoolers exposed to media in early infancy exhibit language delays. In fact, children under two who interact with screens talk later and talk less.[7] They miss nonverbal social cues because they have not been taught to attend to them. Social competence includes learning how to read subtle cues in body language and facial expression.[8] Linguists tell us that very little of a message pertaining to attitudes and feelings is communicated with words. Rather, 38% is communicated through tone of voice and speech pauses, 55% is communicated through body movements and

---

Exploring Video Deficit Effect in 2-Year-Old Children's Playing and Learning with an iPad," Proceedings of the 21st International Conference on Computers in Education 2013. [Accessed May 31, 2019]. http://espace.cdu.edu.au/view/cdu:40222.

[7] Zimmerman et al, Associations between media viewing and language development in children under age 2 years. J Pediatr. 2007;151(4):364–368; H. Duch, E. M. Fisher, I. Ensari, A. Harrington, Screen time use in children under 3 years old: a systematic review of correlates, Int J Behav Nutr Phys Act. 2013 Aug 23;10:102. doi: 10.1186/1479-5868-10-102. Review; G. Geng and L. Disney, "A Case Study: Exploring Video Deficit Effect in 2-Year-Old Children's Playing and Learning with an iPad," Proceedings of the 21st International Conference on Computers in Education 2013. http://espace.cdu.edu.au/view/cdu:40222 [Accessed May 31, 2019]. https://pediatrics.aappublications.org/content/138/5/e20162591#ref-32

[8] R. Pea et al., "Media Use, Face-to-Face Communication, Media Multitasking, and Social Well-Being among 8- to 12-Year-Old Girls," Developmental Psychology 48, no. 2 (March 2012): 327-36, doi:10. I037/a0027030; C. Rowan, "The Impact of Technology on the Developing Child," Huffington Post, May 29, 2013. [Accessed May 31, 2019]. https://www.huffpost.com/entry/technology-children-negative-impact_n_3343245

facial gestures and only 7% comes through actual words.[9]

The American Academy of Pediatrics also found that young children with early exposure to electronic devices have greater difficulty with conflict resolution. Chronically hyper-aroused children become defensive when playing games. They are often unhappy losers, and this behavior shows up, not only online, but on the playground. They exhibit poor frustration tolerance, tearfulness, irritability and, at times, even melt-downs or aggression.[10,11]

Not surprisingly, when these children are in grade school, teachers describe a decline in school performance.[12] This includes the quality of written communi-

---

[9] Mehrabian's Communication Theory: Verbal, Non-Verbal, Body Language. [Accessed May 31, 2019.] https://www.businessballs.com/communication-skills/mehrabians-communication-theory-verbal-non-verbal-body-language/

[10] J.S. Radesky, M. Silverstein, B. Zuckerman, D. A. Christakis. Infant self-regulation and early childhood media exposure. Pediatrics. 2014;133(5). [Accessed May 31, 2019.] https://pediatrics.aappublications.org/content/133/5/e1172. Y. Reid Chassiakos, J. Radesky, D. Christakis, et al., AAP COUNCIL ON COMMUNICATIONS AND MEDIA. Children and Adolescents and Digital Media. Pediatrics. 2016;138(5): E20162593; American Academy of Pediatrics, Council on Communications and Media. Virtual violence statement. Pediatrics. 2016;138(1). https://pediatrics.aappublications.org/content/138/2/e20161298

[11] The American Academy of Pediatrics, Council on Communications and Media. Virtual violence statement. Pediatrics. 2016;138(1). [Accessed May 31, 2019.] https://pediatrics.aappublications.org/content/138/2/e20161298

[12] S. Tang, & M. E. Patrick. (2018). Technology and interactive social media use among 8th and 10th graders in the U.S. and associations with homework and school grades. Computers in

cation.[13] The children themselves describe being bored when reading a book, and some even describe feeling lonely or isolated. Children in classrooms where cell phones are permitted often describe being more interested in looking at their phones than in paying attention to their teacher.

A study examining empathy also found a significant drop in college students' empathy scores after the year 2000. This drop is right in line with the first generation of children who were born into the age of video games and computers.[14]

Common mental health conditions associated with overuse of electronic devices

A modern-day malady affecting two-thirds of all Americans today is chronic sleep deprivation. This is rising in the wake of cell phones, e-readers and other light emitting devices. Over-engagement with screen media impairs the quality of sleep and circadian functioning. This, in turn, is associated with poor memory,

---

Human Behavior, 86, 34-44. dx.doi.org/10.1016/j.chb.2018.04.025.

[13] Y. Reid Chassiakos, J. Radesky, D. Christakis, Zimmerman et al, Associations between media viewing and language development in children under age 2 years. J Pediatr. 2007;151(4):364–368 et al., AAP Council on Communications and Media. Children and Adolescents and Digital Media. Pediatrics. 2016;138(5): E20162593.

[14] S. H. Konrath, E. H. O'Brien, and C. Hsing, "Changes in Dispositional Empathy in American College Students Over Time: A Meta-Analysis," Personality and Social Psychology Review 15, no. 2 (May 1, 2011): 180-98, doi:10.II77/ 1088868310377395.

irritability and impaired school or work performance. The blue spectrum light of the cell phone screen makes the brain think it is still daytime and suppresses the pineal gland from releasing the body's natural sleep aid, melatonin.[15]

Polling of persons using their cell phones revealed that 65% of adults sleep with their phones and nearly all teens do so.[16] Social media is the last thing teenagers check on their phones before going to sleep and the first thing they check when they wake up. Although there are new screens that do not emit blue spectrum light, sleep is still disturbed by staying up late posting pictures on *Snapchat* and *Instagram* and by responding to text messages or *Facebook* posts that come in throughout the night, thereby reducing total sleep time. Recall 75% of 18- to 24-year-olds check their phones in the middle of the night. A meta-analysis on the use of electronic devices among young children demonstrated similar problems with sleep deprivation.[17]

---

[15] R. Salti, R. Tarquini, S. Stagi, et al. Age-dependent association of exposure to television screen with children's urinary melatonin excretion, Neuroendocrinal. Lett. 2006;27(1-2):73–80. [Accessed May 31, 2019] https://www.scientificamerican.com/article/blue-leds-light-up-your-brain/

[16] J.C. Levenson, A. Shensa, J. E. Sidani, J. B. Colditz, B. A. Primack. The association between social media use and sleep disturbance among young adults. Pre. Med. 2016;85:36-41.

[17] E.M. Cespedes, M. W. Gillman, K. Kleinman, S. L. Rifas-Shiman, S. Redline, E. M. Taveras. Television viewing, bedroom television, and sleep duration from infancy to mid-childhood. Pediatrics. 2014;133(5). [Accessed May 31, 2019.] http://pediatrics.aappublications.org/content/133/5/e1163; and Bruni O, Sette S, Fontanesi L, Baiocco R,Laghi F, Baumgartner E.

When teens are asked why they take their phones to bed, they reply using the language of addiction: "I don't know, I just can't help it." Others describe their phones as an extension of their bodies--or even like a lover: "Having my phone closer to me while I am sleeping is a comfort."[18]

The 2017 American Psychological Association study demonstrated that the more people check their electronic devices, the more they describe feeling "stressed out". Millennials described feeling "stressed out" more often than the generations that preceded them.[19] An even higher percentage of adolescents describe themselves as feeling increasingly anxious the more frequently they check their phones.[20] In summary, those who spend more time on their phones accessing social media, gaming, texting and video chatting exhibited more anxiety than those who played sports, engaged in other outdoor physical activities,

---

Technology use and sleep quality in preadolescence and adolescence. J Clin Sleep Med.2015;11(12):1433–1441.

[18] J. M. Twenge, iGen: Why today's super-connected kids are growing up less rebellious, more tolerant, less happy—and completely unprepared for adulthood (New York: ATRIA Books, 2017), p. 50.

[19] American Psychological Association. "Stress in America: The State of our Nation." (Nov. 1, 2017.) [accessed May 31, 2019]. https://www.apa.org/news/press/releases/stress/2017/state-nation.pdf

[20] J. M. Twenge, Have smartphones destroyed a generation? The Atlantic. (Sept. 2017.) [Accessed May 31, 2019.] https://www.theatlantic.com/magazine/archive/2017/09/has-the-smartphone-destroyed-a-generation/534198/

attended religious services and interacted with others face-to-face.[21]

Electronic devices have also spawned a novel 21st century neurosis: "FOMO," or the "fear of missing out."[22] For all the power to link persons day and night, social media has exacerbated the age-old concern, especially among teenagers, about feeling left out. Today's teens spend less time together in person, but when they do congregate, they document their hangouts on Snapchat, Instagram or Facebook, and those who were not invited become keenly aware of this and feel they are missing out.

This fear of missing out is affecting self-esteem and contributing to feelings of inadequacy. Data from surveys are stark: The number of teens who reported "I feel like I can't do anything right" zoomed to an all-time high after 2011 in 8th, 10th and 12th graders.[23] Social media may well play a role in these feelings of inadequacy. We know that teens are more likely to post

---

[21] J. Anderson and L. Rainie. The negatives of digital life. Pew Research Center. [Accessed May 31, 2019]. https://www.pewinternet.org/2018/07/03/the-negatives-of-digital-life/

[22] C.A. Wolniewicz, M. F. Tiamiyu, J. W. Weeks, J. D. Elhai, Problematic smartphone use and relations with negative affect, fear of missing out, and fear of negative and positive evaluation. Psychiatry Res. 2018 Apr;262:618-623; E. Wegmann, U. Oberst, B. Stodt, M. Brand, Online-specific fear of missing out and Internet-use expectancies contribute to symptoms of Internet-communication disorder, Addict Behav Rep. 2017 Apr 14;5:33-42. doi: 10.1016/j.abrep.2017.04.001.

[23] J. M. Twenge, iGen: Why today's super-connected kids are growing up less rebellious, more tolerant, less happy—and completely unprepared for adulthood (New York: ATRIA Books, 2017), p. 100.

only their successes online. Those checking their social media posts do not realize that their friends also fail at things--a fact they would more likely encounter if they spent more time in person with their friends.

How much these devices have changed our global culture in a brief decade! The constant presence of cell phones is likely to affect teens well into adulthood. Depression and anxiety have many causes and too much technology is clearly not the only one. The rates of teenagers, however, experiencing depression and suicide have skyrocketed since 2008.[24] No single factor defines a generation, but the advent of cell phones, which are the teenager's main tool to access their social media, has contributed to a "perfect storm" of a magnitude not seen before. What is at stake is not just how persons experience adolescence, but also how they will experience adulthood. Psychiatrists know that persons who suffer one episode of depression are 50% more likely to become depressed later in life.[25]

Some of these observations in teens and young children have prompted researchers to take a closer look at generational behavior and attitude changes over the past decades.

Persons born between 1965 and 1979, often referred to as Generation X, exhibited the general

---

[24] Center for Disease Control and Prevention. Mortality Data. National Vital Statistics System. [Accessed May 31, 2019]. https://www.cdc.gov/nchs/nvss/deaths.htm

[25] S. Burcusa and W. Iacono. Risk for recurrence in depression. Clinical Psychology Review. 2007 Dec; 27(8): 959–985. Published online 2007 Mar 3. doi: 10.1016/j.cpr.2007.02.005 [Accessed May 31, 2019]. https://www.ncbi.nlm.nih.gov/pmc/articles/PMC2169519/

quality of valuing independence. When in high school, these individuals went to get their drivers' licenses as soon as possible. They even made appointments at the DMV for the day before they turned 16.[26] For Gen X teens, the activity of getting their drivers' licenses allowed them to explore the newfound freedom of escaping the confines of their parents' home and neighborhood.

Not so for iGens, or Generation Z. iGens are the post-millennial generation born between 1997 and 2015. A survey of iGens demonstrated that more than one in four teens does not have a drivers' license at the end of high school.[27] iGens also describe themselves as less likely to leave their home because they are comfortable in their bedrooms.[28] Their social lives are lived on their phones, so they don't need to leave home to spend time with friends. In fact, the number of teenagers who today daily spend face-to-face time together dropped by more than 40% as compared with those who were teenagers around the year 2000.[29] The

---

[26] J. M. Twenge, iGen: Why today's super-connected kids are growing up less rebellious, more tolerant, less happy—and completely unprepared for adulthood (New York: ATRIA Books, 2017), p. 26.

[27] Ibid., p. 28.

[28] D. Rupple, "iGen teens shaped by technology," MasterMedia International (July 12, 2018) [accessed May 31, 2019.] http://www.mastermediaintl.org/igen-teens-shaped-by-technology/

[29] Social media causes isolation in teens. Roanoke Area Youth Substance Abuse Coalition. [Accessed May 31, 2019.] https://raysac.org/2018/03/20/social-media-causes-isolation-in-teens/

roller rinks, miniature golf courses, town pools and local necking spots have all been replaced by virtual spaces.

iGens are the generation who most often frequent social networking sites and paradoxically most often agree with the statement, “A lot of times I feel lonely” or “I often feel left out of things” and “I often wish I had more good friends.”[30]

Teens’ feelings of loneliness spiked in 2013 and have remained high ever since.[31] Again this does not prove that teens who spend more time online are lonelier than teens who spend less time online. However, as compared with millennials and older generations, loneliness is described as a more common experience by iGens. This is ascribed to be, in part, due to their less frequent face-to-face interactions and subsequent increased experiences of FOMO.

---

[30] V. Manning-Schaffel, Americans are lonelier than ever, but GenZ may be the loneliest. NBC News: Better. (May 14, 2018.) [Accessed May 14, 2018.] https://www.nbcnews.com/better/pop-culture/americans-are-lonelier-ever-gen-z-may-be-loneliest-ncna873101

[31] J. Ducharme, Depression has spiked by 33% in the last five years, a new report says. Time. (May 10, 2018) [Accessed May 31, 2019.] http://time.com/5271244/major-depression-diagnosis-spike/

More concerning mental health conditions related to overuse of cell phones

Sexting

Concerning behaviors with potentially more harmful consequences of cell phone overuse includes sexting.[32] It is estimated that 22% of teenage girls and 18% of teenage boys have sent or posted a nude or semi-nude picture or video of themselves.[33] Unfortunately, with maturing age, these numbers increase significantly: 36% of females between the ages of 20 and 26 have sent or posted a nude or semi-nude picture or video of themselves and 31% of adult males have done so.[34] This trend indicates that with maturing chronological age, young adults of today are sexting more often than in their teenage years.

Teens and young adults may not grasp the consequences of their actions. Sexting becomes a legal issue when teenagers under age 18 are involved because any nude photos they may send of themselves place the recipients in possession of child pornography.

---

[32] J. R. Temple, V. D. Le, P. van den Berg, Y. Ling, J.A. Paul, B.W. Temple. Brief report: Teen sexting and psychosocial health. J Adolesc. 2014;37(1):33–36.

[33] Sex and tech: Results from a survey of teens and young adults. Cosmogirl.com. [Accessed May 31, 2019.] https://powertodecide.org/sites/default/files/resources/primary-download/sex-and-tech.pdf

[34] Sexting statistics. Statistic Brain Research Institute. (April 26, 2017.) [Accessed May 31, 2019.] https://www.statisticbrain.com/sexting-statistics/

Consider this example:[35] The case began when an eighth-grade girl at the Montessori Middle School of Kentucky was suffering from anorexia and had to relocate to a treatment center in Arizona. After two months at the Remuda Ranch facility, she returned to Kentucky to finish the school year.

Soon after she returned home, now age 14, she developed a crush on a boy, age 14. The two would attend the same Lexington Catholic High School in the fall as freshmen. According to the girl, the boy soon "made several telephone calls to the girl telling her that he wanted her to create a video with her telephone showing her 'pleasuring herself', a video which the boy said he would use when he masturbated." The girl refused at first, but the boy told her that "he would not be her friend at Lexington Catholic High School" without the video.

This continued for some time, with the boy sending text messages to the girl in which he promised to "keep the sexually explicit video secret." The girl gave in. According to the girl, she was "finally coerced, enticed, and persuaded" to produce an 8- to 10-second video clip of herself masturbating, which she sent to the boy using her cell phone. Only a few weeks into the school year, the boy was convinced by one of his friends to transfer the cell phone video to his computer. From there, it was a small step to uploading the short clip onto the Internet, with predictable results.

---

[35] N. Anderson, 14-year old child pornographers? Sexting lawsuits get serious. Ars Technica (March 5, 2011) [Accessed May 31, 2019.) https://arstechnica.com/tech-policy/2011/03/14-year-old-child-pornographers-sexting-lawsuits-get-serious/

The video was "uploaded by Lexington Catholic High School students to their iPods in order to share the sexually explicit video with as many students at Lexington Catholic high school as possible." Students at three other local high schools viewed the clip as well, and students began calling the girl in the video "a whore, 'nasty Nat,' and the 'porn queen'."

The girl's mother went to the police. Nothing came of the investigation because the girl "could be seen as having been as guilty as the boy for producing and distributing child pornography."

The girl claimed she was "subjected to a daily routine of harassment" at school and suffered from "depression, anxiety, mental anguish, embarrassment, and extreme stress." She was eventually advised, for her own mental and physical well-being, to finish high school elsewhere.

The tale is undeniably horrific. Even more so when one considers the girl's existing psychological vulnerability. But did someone break the law? If so, what law was it?

The chief federal judge for the Eastern District of Kentucky (Judge Jennifer Coffman) stated: "[P]rosecutors have begun to charge minors under child pornography statutes in sexting cases around the country, based on the sexual exploitation law that applies to all persons regardless of age or social equality. For these reasons, 14-year-olds can in fact sexually exploit other children and can be held liable for that exploitation."

Civil authorities are wrestling with the complexity of sexting and when the law should get involved in such cases. Consider the following (cited by Judge Coffman who presided at the above court hearing):

- "In Massachusetts, police were considering criminal charges after an eighth-grade girl sent pictures of herself to her eighth-grade boyfriend, who then sold the images to other children for $5."
- "In Virginia, two male high school students, aged 15 and 18, were charged with possession of child pornography and electronic solicitation for nude and semi-nude images of minor girls contained on cell phones, which they traded among themselves."
- In Florida, "[A]n 18-year-old high school senior who had recently broken up with his 16-year-old girlfriend e-mailed everyone on his ex-girlfriend's e-mail contact list nude images that she had originally e-mailed only to him. He was convicted under state child pornography law and required to register as a sex offender."
- "In Wisconsin, [a] 17-year-old boy was charged with possession of child pornography after posting naked pictures of his 16-year-old ex-girlfriend online with crude comments."
- "Also in Wisconsin, an 18-year-old was sentenced to 15 years in prison for an extortion scheme in which he tricked male classmates into sending him nude photos of themselves and then blackmailed them with exposure if they refused to have sex with him."

Clearly, sexting is complicated when the images hit the internet, or when extortion, coercion, and payment are involved. Regardless of their original intent, such pictures and videos may provide fodder for those attracted to child pornography.

Once on the internet, any and all material is almost impossible to remove. Sexting, texting and other such electronic behaviors can also haunt a college applicant or prospective employee years later. More colleges are checking online profiles looking for an applicant's suitability or giant red flags about poor judgment.

Cyberbullying

Cyberbullying is also on the increase. Twenty to 40% of adolescents report having been victims of cyberbullying and 25% of them repeatedly.[36] Teens who are cyberbullied often say that it is nearly impossible to get away from their tormentors. Unlike bullies who present themselves in person, it can be impossible to avoid a cyber-tormenter unless teenagers give up their phones entirely, something they typically are not willing to do.[37] More than 50% of adolescents do not tell their parents that they are

[36] S. Cook, Cyberbullying facts and statistics 2016-2018. Comparatech. (November 12, 2018.) [Accessed May 31, 2019.] https://www.comparitech.com/internet-providers/cyberbullying-statistics/

[37] J. Raskauskas, A. D. Stoltz, Involvement in traditional and electronic bullying among adolescents. Dev Psychol. 2007;43(3):564–575.

victims of cyberbullying.[38] Many of these teens not only suffer anxiety and depression, but are also at greater risk of self-harm and suicidal behaviors.[39]

Sex-trafficking

Social media has also increased the incidence of child sex trafficking. Predators gain access to children and teens through social networking, chat rooms, email and online games. Most victims are between the ages of 13 - 17.[40] Sex trafficking is a growing industry behind drugs and weapons.[41] It is important to educate teenagers about protecting themselves from exploitation. *Backpage* has been a frequent site for

---

[38] S.K. Schneider, L. O'Donnell, A. Stueve, R.W. Coulter. Cyberbullying, school bullying, and psychological distress: a regional census of high school students. Am J Public Health. 2012;102(1):171–177. McDougall P, Vaillancourt T. Long-term adult outcomes of peer victimization in childhood and adolescence: Pathways to adjustment and maladjustment. Am Psychol. 2015;70(4):300–310.

[39] A. John, A. C. Glendenning, A. Marchant, P. Montgomery, A. Stewart, S. Wood, K. Lloyd, and K. Hawton. Self-harm, suicidal behaviors and cyberbullying in children and young people: systematic review. J Med Internet Res. 2018 Apr 19;20(4).

[40] T. Swarens. Who buys a trafficked child for sex? Otherwise ordinary men. USA Today Network. [Accessed May 31, 2019.] https://www.usatoday.com/story/opinion/nation-now/2018/01/30/sex-trafficking-column/1073459001/

[41] J. Greenberg, Yes, human trafficking ranks no. 3 in world crime. Politifact. (July 26, 2016.) [Accessed May 31, 2019.] https://www.politifact.com/truth-o-meter/statements/2016/jul/26/amy-klobuchar/yes-human-trafficking-ranks-3-world-crime/

sex-trafficking and a difficult one to eliminate from cyberspace.

In April, 2018, Ferrer, the CEO of *Backpage* finally admitted to conspiracy, money laundering and facilitating prostitution. He received a five-year jail sentence.[42] During the same period, President Trump signed two bills into law to fight online sex trafficking.[43]

---

[42] U.S. Department of Justice, Backpage's co-founder and CEO, As well as several backpage-related corporate entities, enter guilty pleas. Justice News. (April 12, 2018.) [Accessed May 31, 2019.] https://www.justice.gov/opa/pr/backpage-s-co-founder-and-ceo-well-several-backpage-related-corporate-entities-enter-guilty

[43] H.R.1865 - Allow states and victims to fight online sex trafficking act of 2017. (April 11, 2018.) [Accessed May 31, 2019.] https://www.congress.gov/bill/115th-congress/house-bill/1865

Section 2

Screen Addiction

Is there such a thing?

In the 1950s, two neuropsychologists, Peter Milner and James Olds, accidently discovered that implanting electrodes in a specific part of the brain of a rat and coupling the electrode with a metal bar resulted in the rat pressing the bar more than 7,000 times in twelve hours. The rat did not eat nor engage with other mates. This repetitive behavior ceased only when the rat died of exhaustion. Further investigation by Milner and Olds resulted in the discovery that the part of the brain in which the electrode was placed regulated pleasure. They later dubbed this part of the brain "the pleasure center".[44] An obvious question for neuroscientists and addiction specialists was, "Can certain human behaviors trigger the pleasure center and create an

[44] J. Olds, & P. Milner. (1954). Positive reinforcement produced by electrical stimulation of septal area and other regions of rat brain. Journal of Comparative and Physiological Psychology, 47(6), 419-427.

addiction?" With the advent of high-speed internet, cell phone and social media, the answer to this question has become a resounding "yes". So long as a behavior stimulates the reward center and eases physiological discomfort, an addictive response is likely to ensue. Neuroscientists now accept that an addiction can be triggered by the action of certain substances that are consumed or by the action of certain behaviors that are repeated.

This finding contributed to conceptualizing addiction, not only as a disease, but as a sort of a learning disorder which arises when the pleasure center is over-stimulated by certain actions which relieve discomfort. Whether a person uses an opioid to relieve discomfort or compulsively checks social media on a cell phone, if the pleasure circuits of the brain are stimulated, addiction can be the result.[45]

Before we go any further let's review some basic neurophysiology. Our brain is jam packed with over 100 billion cells. Each cell has extensions or wires connecting other cells allowing them to communicate and receive messages from one another across synapses or connection sites. Scientists estimate that the number of nerve cell connections contributes to 100 trillion operations per second![46] The different

---

[45] Y.H. Lin et al, Incorporation of Mobile Application (App) Measures Into the Diagnosis of Smartphone Addiction, J Clin Psychiatry. 2017 Jul;78 (7):866-872. doi: 10.4088/JCP.15m10310.

[46] C. Zimmer, 100 trillion connections: New efforts probe and map the brain's detailed architecture. Scientific American. (January 2011.) [Accessed May 31, 2019]. https://www.scientificamerican.com/article/100-trillion-connections/

masses of cells have different specific functions. The prefrontal lobes, located above the eyes, are responsible for thinking, reasoning and planning. Adults can provide learning moments to strengthen these skills in children. The skills of the prefrontal lobes, therefore, are learned. Surrounding each nerve is the myelin sheath. It is like the plastic around an electrical cord which guides the neurotransmitters to travel along the nerve. Neuroscientists refer to this as the "use it or lose it" principle.[47] Brain cells are like muscles. Their connections become stronger the more they are used. That is, the connections that are most frequently used become more efficient and the connections that are not used are pruned back or removed from the developing brain.

One of the neurotransmitters that travels along the nerve cells in the brain is called dopamine. The brain needs dopamine to function properly as it influences the pleasure center of the brain. The more dopamine is released the more pleasure as experienced, at least to a point.

Michael Zieler, a psychologist, published his results with pigeons demonstrating that far more dopamine is released when the reward for food pellets came at unpredictable times.[48] Social media designers, whose work includes refining technology, run thou-

---

[47] K. Wong, Use it or lose it. Scientific American. (October 16, 2000.) [Accessed May 31, 2019]. https://www.scientificamerican.com/article/use-it-or-lose-it/

[48] M. D. Zeiler, and A. E. Price, "Discrimination with variable interval and continuous reinforcement schedules", Psychonomic Science January 1965, Volume 3, Issue 1–12, pp 299–300.

sands of tests on millions of users to learn which tweaks work to increase the use of a social media app, for example, and which do not. A team of *Facebook* web developers did a similar experiment on hundreds of millions of users. *Facebook* already had 200 million users when they tested the "Like" button.[49] This was the first of many more feedback buttons now available on *Facebook*. The "Like" button completely changed how *Facebook* was accessed. *Facebook* began as a passive way to track friends' lives but evolved into an interactive social media app with the same unpredictable feedback trigger that motivated Zieler's pigeons. Users were gambling every time they shared a photo, web link or status update. A post with zero "Likes" was a kind of public humiliation. *Twitter, Instagram, Google* and *YouTube* are just a few of the social networking sites that have also harnessed the power of unpredictable positive feedback (a.k.a., variable reinforcement) to trigger the release of dopamine.

Now factor in something called "operant conditioning" or "cueing". A chirp that tells you about a new *Tweet* or a ding that informs you about a new *Facebook* post or email conditions you to anticipate a response before you even know what the message contains. These noises act as a stimulus and condition you to check your email, post online or perform some

---

[49] G. Moss, How the Facebook "Like" button has changed through the years. Bustle. (February 4, 2016.) [Accessed May 31, 2019]. https://www.bustle.com/articles/139394-how-the-facebook-like-button-has-changed-through-the-years

other "screen time" interaction. This operant conditioning of sounds increases the addictive effect.

Addictive behaviors have been around for a long time. With the advent of high-speed internet and smaller, carry in your pocket or purse devices, it is harder to resist not engaging with them since we carry them almost everywhere we go.

Withdrawal and craving from screen addiction can be so strong that just sitting in front of a computer or merely opening up the internet on your cell phone or tablet can trigger a release of dopamine that makes you seek more social media, even though you receive less pleasure from doing so. What started as pleasant screen time has become an addictive activity to keep you from feeling restless.

Teenagers have a highly neuroplastic brain and are vulnerable to the neurochemical changes that digital devices can create in the brain from hyperarousal. Excess of screen time too early in life can be damaging and addicting. This is especially true during childhood because the brain is still developing–a process that isn't complete until a person is around 25 years old. We might ask when did Steve Jobs give his children iPads? The answer? Never![50] What about Evan Williams, the founder of Blogger, Twitter and Medium? He bought hundreds of books for his two young sons and refused

---

[50] M. Duczeminski, Even Steve Jobs didn't let his kids use iPads: Why you should limit technology use for your kids. Lifehack. [Accessed May 31, 2019.] https://www.lifehack.org/299415/even-steve-jobs-didnt-let-his-kids-use-ipads-why-you-should-limit-technology-use-for-your-kids

to give them an iPad.[51] Recommendations to help prevent or address the addictive effects of electronic devices is outlined at the end of this section.

Online gaming

Researchers agree that video games can be helpful. For example, educational games can boost learning, and action games can improve vision and spatial skills. Video games have also been used successfully to teach kids self-care skills for asthma and diabetes. However, video games have also been detrimental to learning, socialization and health when engaged excessively.

Surveys tell us that the videogame market brought in $90 billion in revenue in 2017 and this number increases each year.[52] The US pew survey cites that 90% of 12- to 17-year-old American teens and 60% of Americans between the ages of 18 and 29 play video games.[53] Puzzle and strategy games are among the most popular types of video games.[54] Also, online

---

[51] A. Alter, Tech bigwigs know how addictive their products are. Why don't the rest of us? Wired. (March 24, 2017.) [Accessed May 31, 2019]. https://www.wired.com/2017/03/irresistible-the-rise-of-addictive-technology-and-the-business-of-keeping-us-hooked/

[52] 2019 video game industry statistics, trends & data. WePC. (April 2019.) [Accessed May 31, 2019]. https://www.wepc.com/news/video-game-statistics/

[53] A. Perrin, 5 facts about Americans and video games. Pew Research Center. (September 17, 2018.) [Accessed May 31, 2019]. https://www.pewresearch.org/fact-tank/2018/09/17/5-facts-about-americans-and-video-games/

[54] Ibid.

gaming is available on almost any electronic platform today, making online games available 24 hours a day, 7 days a week.

One of the more popular massively multiplayer online role-playing games is *World of Warcraft*.[55] *World of Warcraft* took second place among the best-selling PC games of all time around the world, with a total of 6.33 million units sold as of January 2018.[56] If *World of Warcraft* players formed a nation, they would be the twelfth largest nation in the world. Players of *World of Warcraft* create avatars and join guilds where a small band of players shares resources and chats regularly in their guild specific chat rooms. Players describe it being hard to sleep at night when they know that several of their guild mates in Japan, Slovakia and Germany, for example, are on an epic quest without them. Not surprisingly, online gaming contributes to decreased hours of sleep, doing less homework and missing classes the next day. Guild mates can become closest friends, which players describe as lacking in their offline world. For some

[55] Distribution of World of Warcraft characters in U.S. and EU realms as of February 2019, by class. Statista. [Accessed May 31, 2019]. https://www.statista.com/statistics/276318/distribution-of-world-of-warcraft-characters-by-class/

[56] 2019 video game industry statistics, trends & data. WePC. (April 2019.) [Accessed May 31, 2019]. https://www.wepc.com/news/video-game-statistics/

players, *World of Warcraft* becomes their predominant means of socializing.[57]

Excessive online gaming can also present serious health risks in some individuals. There are examples of teenagers who have played online games, sitting for hours in a row, getting little to no sleep and not taking time out to eat or drink sufficient fluids. A Taiwanese teenager was found dead after sitting for 40 hours in an internet cafe playing *Diablo 3*. At the time, doctors speculated he died from a heart attack caused by a blood clot that formed during the long session.[58] A South Korean 28-year-old collapsed after playing *StarCraft* for 50 hours with brief pauses only to go to the toilet and naps. It was thought that he died from heart failure secondary to exhaustion. He was not known to have prior health problems. He had been terminated from his job for missing work to play computer games.[59] Ten percent of online game players

---

[57] H. Longman, E. O'Connor, P. Obst. The effect of social support derived from World of Warcraft on negative psychological symptoms. Cyberpsychol Behav. 2009; 12(5)z;563-566. Doi:10.1089cpb.2009.0001); Zanetta Dauriat F. Motivations to play specifically predict excessive involvement in massively multiplayer online role-playing games: evidence from an online survey, Eur Addict Res. 2011;17(4):185-9. doi: 10.1159/000326070.

[58] C. Sieczkowski, Diablo 3 death: Teen dies after playing game for 40 hours straight. Huffpost. (July 18, 2012.) [Accessed May 31, 2019]. https://www.huffpost.com/entry/diablo-3-death-chuang-taiwan_n_1683036

[59] StarCraft Players, Lee Seung Seop. Fandom. [Accessed May 31, 2019.] https://starcraft.fandom.com/wiki/Lee_Seung_Seop

admit to playing an average of 63 hours per week and 44% of players consider themselves addicted.[60]

While these are extreme cases, they are a reminder that sitting at a computer, game console or on a cell phone for days, whether to play *World of Warcraft* or other games, is not healthy for anyone.

These findings contributed to a new disorder being proposed in the most recent DSM V (*Diagnostic and Statistical Manual of Mental Disorders*, Fifth Edition) called Internet Gaming Disorder.[61] Internet gaming disorder exhibits symptoms like substance-abuse disorders including tolerance, withdrawal and craving. The specific criteria are outlined in the figure below.

Figure 1. Proposed DSM-5 Criteria for Internet Gaming Disorder

Persistent and recurrent use of the Internet to engage in games, often with other players, leading to clinically significant impairment or distress as indicated by 5 (or more) of the following in a 12-month period:

1. Preoccupation with Internet games. (The individual thinks about previous gaming activity or anticipates playing the next game; Internet gaming becomes the dominant activity in daily life.)
2. Withdrawal symptoms when Internet gaming is taken away. (These symptoms typically are described as

[60] B. Conrad, 44% of World of Warcraft players say that they are addicted to the game. Techaddiction. (2013.) [Accessed May 31, 2019.] http://www.techaddiction.ca/44-percent-world-of-warcraft-gamers-addicted.html

[61] American Psychiatric Association, Diagnostic and Statistical Manual of Mental Disorders, 5th ed. (Arlington, VA: American Psychiatric Association, 2013), p. 795-6.

irritability, anxiety, or sadness, but there are no physical signs of pharmacologic withdrawal.)
3. Tolerance-the need to spend increasing amounts of time engaged in Internet games.
4. Unsuccessful attempts to control the participation in Internet games.
5. Loss of interests in previous hobbits and entertainment as a result of, and with the exception of, Internet games.
6. Continued excessive use of Internet games despite knowledge of psychosocial problems.
7. Has deceived family members, therapists, or others regarding the amount of Internet gaming.
8. Use of Internet games to escape or relieve a negative mood (e.g., feelings of helplessness, guilt, anxiety).
9. Has jeopardized or lost a significant relationship, job, or educational or career opportunity because of participation in Internet games.

There is also a growing concern about the violent content of video games. 97% of youths ages 12 through 17 play some kind of video game, and 2/3 of teens play action and adventure games that contain violent content.[62] The online game *Grand Theft Auto* is one example. Its popularity is reflected in the Guinness Book of World Records for seven record-breaking achievements in 2013 including the best-selling video game in twenty-four hours and the fastest video game to gross one billion dollars.[63] The game contains scenes

[62] M. Anderson and J. Jiang, Teens, social media, & technology, 2018. Pew Research Center. (May 31, 2018.) [Accessed May 31, 2019.] https://www.pewinternet.org/2018/05/31/teens-social-media-technology-2018/

[63] L. Karmali, GTA 5 currently holds seven Guinness World Records. IGN. (October 9, 2013.) [Accessed May 31, 2019.]

of gruesome waterboarding torture in which the player must choose to continually inflict pain if he wants to advance to the next level. Women are frequently depicted as sexual objects which include fondling. Players also have the opportunity to make their avatars drink alcohol and use other drugs.[64] As graphic as the violence is on TV and in movies, in the video game medium, the player, not the actor on the screen, controls the action. In short, players win more points by engaging in criminal behaviors. Parents also need to know that this game involves chatting with unknown players.

Children who excessively play violent video games may skip meals and instead live off snacks and energy drinks to keep awake so they can keep playing. They are not solving and negotiating conflicts with their peers. They are missing opportunities to gain needed cooperative learning and social skills in the real world. They can become even more inept at communicating cooperatively with actual people and instead can lose themselves in video games that give them feelings of control, mastery and exhilaration so that the real world of slowness and struggle, decisions and demands becomes less appealing. In contrast to the duration of other forms of entertainment, parents may rent a video for a day, then return it. A family may go to Disneyland

---

https://www.ign.com/articles/2013/10/09/gta-5-currently-holds-seven-guinness-world-records

[64] A. Hern, Grand Theft Auto 5 under fire for graphic torture scene. The Guardian (September 18, 2013.) [Accessed May 31, 2019.] http://theguardian.com/technology/2013/sep/18/grand-theft-auto-5-under-fire-for-graphic-torture-scene

for a couple of days, then return home. Nobody keeps going daily as a tourist.

Fallout 3, another top selling Internet game, includes "realistic dismemberment," such as slow motion decapitation, while players use a kitchen knife to kill family members and pets.[65,66] *Diablo III* depicts battles accompanied by screams of pain and blood splattering effects.[67] In *Postal*, another online game, players receive points for killing as many innocent victims as possible while they begged for mercy.[68]

Ever since the first online video games went on the market, the advertising focus has been on the realism of the displays. In fact, reviews of new violent video games boast about how well they depict the violence. It is unnerving that we have come to accept such depictions as acceptable recreation for children and teenagers.

There is not any simple lesson regarding the effects of video games on children, however. There are clearly both costs and benefits associated with video games, and these vary depending on how and when children

---

[65] M. B. Sauter, The seven best-selling violent video games. 24/7 Wall St. (March 14, 2013.) [accessed May 31, 2019.] https://247wallst.com/special-report/2013/03/14/the-seven-best-selling-violent-video-games/2/

[66] Fallout 3, Entertainment Software Rating Board [Accessed May 31, 2019.] http://www.esrb.org/ratings/Synopsis.aspx?Certificate=25545&Title=Fallout+3

[67] Diablo 3, Entertainment Software Rating Board [Accessed May 31, 2019.] http://www.esrb.org/ratings/synopsis.jsp?Certificate=31460

[68] Postal, Ripcord Games. [Accessed May 31, 2019.] https://www.amazon.com/Postal-Pc/dp/B000AAXD9A

play. It is also possible that the effects vary with content of a game. Dr. Donald Shifrin, of the American Academy of Pediatrics, stated that when youngsters engage in video games the object is excitement.[69] Children and teens discover that playing online games is exhilarating. They seek more excitement by playing more online games. Over time, they need to seek increased levels of excitement in order to receive the same rush from playing online games. When more and more stimulation becomes necessary to achieve feelings of excitement, the player has moved into the stage addiction specialists refer to as habituation.[70] Repetitively practicing violence while in high states of arousal, and deriving psychological satisfaction from this practice, may contribute to a conditioning process toward violence.[71] That is, a child's brain may wire neurological circuits that associate feelings of satisfaction with inflicting

---

[69]Dave Grossman and Gloria Degaetano Stop Teaching Our Kids to Kill : A Call to Action Against TV, Movie and Video Game Violence, Harmony Books, New York, (2014), p. 86-87.

[70] K. Bailey, R. West, and C. A. Anderson. 2010. A negative association between video game experience and proactive cognitive control. Psychophysiology. 47(1):34-42. D. Bavelier, C. S. Green, D. H. Han, P. F. Renshaw, M. M. Merzenich, and D. A. Gentile. 2012. Brains on video games. Nat Rev Neurosci. 2011 Nov 18;12(12):763-8. doi: 10.1038/nrn3135. D. Grossman and G. Degaetano, Stop teaching our kids to kill: A call to action against TV, movie and video game violence (New York: Harmony Books, 2014), p. 86-87.

[71] The American Academy of Pediatrics, Council on Communications and Media. Virtual violence statement. Pediatrics. 2016;138(1). [Accessed May 31, 2019.] https://pediatrics.aappublications.org/content/138/2/e20161298

pain on others. With more gaming, the child can become desensitized, that is develop a comfort level with inflicting pain on other people in a game. With desensitization, persons become numb or dull to what they are doing, feeling or thinking. Killing, maiming and hurting others is rehearsed in countless ways, hundreds of times a day in a game. Killing in the game is cued with the rush of winning points and more points for head shots in some games. This is called stimulus-response training.

Law-enforcement and the military use stimulus-response techniques when training individuals to shoot or not shoot. These adults have a mature brain as well as ongoing professional training. The stimulus-response training is much more dangerous in the hands of children who lack the maturity to discern the important distinction between virtual and actual reality. A game was even made of the horrifying Columbine massacre.[72] In many shooting sprees, the shooters put on a costume, such as a trench coat or military uniform, copying the video game characters. When children play an interactive video game there is always the intent to shoot. To keep the excitement high and the heart rate up, the thinking functions are repressed. Dr. Brandon Centerwall in the documentary *Healthy Brain Development in a Media Age* stated, "Since youngsters do not have the brain capacity yet for analysis, evaluation, or moral judgment, they are developmentally unable to discern

---

[72] Super Columbine Massacre RPG. [Accessed May 31, 2019.] http://www.columbinegame.com/

the difference between fantasy and reality; if they did we wouldn't have so many children believing in Santa Claus or the tooth fairy. Therefore, they are incapable of interpreting violent images, of making personal sense out of them."[73] Pediatricians, mental health care providers and FBI consultants tell us that teens and young persons involved in actual shooting sprees often dress up like their virtual world counterparts as if to more closely identify with them, thereby enabling the shooters to "follow the script" they rehearsed daily in their video game play.[74]

Play is intended to be a process of learning and practicing appropriate social roles. Aggressive play is part of a normal young child's experimentation. In fact, parents who refused to buy toy guns for their young sons report that guns are made from sticks, carrots, bananas or toilet rolls. I even heard of a mother who had banned guns from her home complaining that her young boys took a peanut butter sandwich, rolled it up and made a toy gun out of it. Children wanting to act out what they see on TV is a normal process. However, exclusively watching and playing violent interactive online roles repeatedly, such that these images become the only source of a child's play fantasies, is likely to result in a child's personality imprinting on poor role models.

---

[73] D. Grossman and G. Degaetano. Stop Teaching Our Kids to Kill : A Call to Action Against TV, Movie and Video Game Violence. Revised and Updated Edition. (New York: Harmony Books, 2014), p. 115.

[74] Ibid., p. 127-128.

A controversy has developed over the relationship between violent video game play and aggression. Dr. Prescott and colleagues from the Department of Psychological and Brain Sciences at Dartmouth College wrote, "Whereas the majority of those who conduct research on this topic argue that playing such games increases aggressive behavior, a vocal minority has argued that the relation of game play and real-world aggressive behavior is at best overstated and at worst spurious." To better understand the link between video game violence and aggressive behavior, they performed a meta-analysis of 24 prospective studies encompassing more than 17,000 participants aged 9 through 19 years. The results showed that ethnicity was a statistically significant moderator for the fixed-effects models ($P < .01$), with the largest effect observed among Caucasians, an intermediate effect among Asians, and a nonsignificant effect among Hispanics. James D. Sargent, MD, from the Department of Pediatrics at Geisel School of Medicine, Dartmouth College, stated, "I hope our findings prompt skeptics to reevaluate their position, especially since some of our other research indicates that violent video game play may increase deviance with implications for multiple risk behaviors."[75]

We need to provide counter scripts and role models that allow children, at every stage of development, to construct stories about themselves as

[75] Prescott AT, et al., Meta-analysis of the relationship between violent video game play and physical aggression over time, October 2, 2018,115 (40) 9882-9888 Proc Natl Acad Sci. 2018;doi:10.1073/pnas.1611617114.

law abiding empathic socially competent and emotionally intelligent individuals. Yes, we can feel like salmon swimming upstream against the current, and a strong one at that. Yes, it will take more than the individual energy, courage and stamina that we feel we have. Yes, it will mean consciously taking time to participate with others in schools and communities to weave collectively a protective safety net for children. In other words, it will take a lot of energy from a lot of people. However, the activities do not have to be time-consuming. Teaching children about virtue through example, responsibly maintaining an active presence in their lives and praying for our children can be a powerful means of achieving our goal.

In summary, a few recommendations:

- Become educated about the basics of brain development in the early years and how important hands-on, unstructured and social play is to building language, cognition and social skills
- Choose educational content that is developmentally appropriate for children
- Play 15 minutes of the game and find out what video games and apps children are playing
- Read reviews of the game and app
- Guide children about how to maintain safety and privacy online
- Help young children begin to REFLECT on the effect media violence has on them:

    - Was it real or pretend? How do you know?
    - What would happen to you if someone really hurt you, your mom or your dad like that?
    - How much do you think that punching, shooting or stabbing would hurt that person?
    - What would have been a better way to solve the problem than fighting, killing, etc.?
    - Let's talk about it
    - Remember, addressing media violence as a family takes time; it is not an event, it is an ongoing learning process.

- Address the "I'm bored" reaction

    - Rather than give suggestions to resolve kids' feelings of being "bored" that they will likely reject, introduce "think-time-outs". That is, let them try to figure out an alternative for themselves. For example, a mother with whom I worked suggested to her son, "Sit on the couch and stay bored. Yes, that's right, just sit there," she told him. "When you are through being bored, you can get up and play with one of your toys in your room." After sitting for a while, he got up off the couch to play Legos.
    - What about us adults? We can equate stillness to boredom. Actually, stillness gives the mind the space it needs to be creative, to come up with new ideas. Consider choosing situations in which you would usually pull

out your cell phone such as in an elevator, waiting in line, or having lunch. Instead, notice what is around you, taste the food you are eating or look at the sky. By practicing being attentive to your surroundings, you are building muscles of attention and the ability to ignore distractions.

- Place computers, cell phones, and gaming devices in a central location and not behind closed doors. Doing so allows parents to monitor kids' activities to help protect them.
- Set and enforce limits on screen time
    - Children are often unable to accurately judge the amount of time spent on electronic devices. Further, they are unconsciously reinforced to keep using them.
    - Most of the time a child spends on electronics has nothing to do with academics.
- Establish media guidelines and follow them consistently
    - Communicate these media guidelines to other caregivers, such as babysitters and grandparents
    - Have children complete their homework before you allow them to play a video game or watch TV

  - Use a timer to indicate when screen time should end

    - Create weekly time coupons. For example: Ten 30-minute coupons for young children and ten 1-hour coupons for older kids. Give the children all the coupons at the beginning of the week and let them choose when they will "redeem" the coupons. This is an excellent way to teach children self-control and choice making.

  - Create "unplugged" spaces and times in the home

    - Consider the importance of not displacing sleep, exercise and social interactions with electronic media.
    - Recommend not using electronic media during meals, in bed and certainly not while driving

- Consider cell phone, tablet and computer Internet firewalls and filtering systems as appropriate for individuals or families. There are many on the market. Some suggestions include:

  - *Circle Go,* which allows parents to manage all home-connected devices, including personal computers, mobile phones and PlayStations.

- Parents choose specific apps or platforms to allow or block, for example, Amazon, Facebook, HBO, YouTube, etc.
- Parents can also

  - allow or block explicit content categories such as dating, social media, etc.
  - set limits for time usage, for example, 2 hours of Internet per day
  - configure time limit per app such as 15 minutes on Facebook
  - block all apps and usage at bedtime
  - track minutes of usage per app or sites accessed

- *Mobicip* and *Covenant Eyes* are also popular Internet filtering and firewall systems
- Other app-blocking devices include:

  - *Freedom* for Apple and Windows products
  - *Offtime* for Android products
  - *GizmoPal* is a GPS tracking watch for kids that is useful for some families

- For parents with young children, the American Academy of Pediatrics recommends limiting the use of technology to one hour per day for children 2-5 years of age. For older children, create a balanced media plan consistent with family values.

    - This is not easy to do in today's world, acknowledging that some screen time is all but inevitable. Instead of banning screen time all together, encourage:

        - Children to connect what they see in the screen world to their experience of the real world
        - Children to engage in real activities in preference to virtual ones. If an app asks children to sort wooden blocks by color, for instance, parents might ask children to sort real wooden blocks by color. Parents might also ask children to sort their clothes by color

    - No experience should be confined to the virtual world of electronic devices that mimic reality. Bridging screen to reality experiences improves learning because it requires children to repeat what they've learned and encourages them to generalize what they've learned beyond a single situation.

We know the tricks that advertisers use to tempt us—a mouth-watering sundae or a shiny red car. Electronic design engineers also know how to get our attention. The color red in notifications on our cell phones and tablets triggers a desire to see what it is all about. Auto-play in videos can get us hooked to watching much longer than we intended. More and more cell phone and tablet functions are designed to get us to spend more time using these devices.

Here are a few more suggestions to help you monitor your tech usage:

- Say no to notifications. A notification is intended to let you know something important needs attention. Most cell phone and tablet notifications are machine-automated and do not involve people. They are intended to get you to engage with an app you might not have otherwise thought about. Set notifications for things that are important to you.
    - For example, you might want notifications when you receive a message from your child, spouse or supervisor.
- Put the cell phone on "do not disturb" mode and allow only calls and messages from "favorites" to whom you have a responsibility. You might be surprised how much time interruptions from messages take away from what is important to you.

- Remove social media and other apps from the cell phone. Tapping on an app, such as *Facebook* or *Twitter*, is easy to do—without much thinking. Limit your access to these apps to your home computer. If you have the urge to check your social media more often, ask yourself why: are you bored, lonely, anxious, bothered or concerned? See if this need could be addressed more effectively in a real-life interaction or activity.
- Keep only the apps on the home screen that are necessary. For example, some important apps to keep on your home screen may be your contacts, text messaging, email, maps and calendar apps that you use daily for family, work and select friends. Move your other apps to your second and third screens. What you do not see right away when you pick up your phone or tablet will be less tempting to access.
- Help replace screen time with positive face-to-face activities

    - Consider scheduling a range of activities, as well as one-on-one time with friends and family that communicate "I want to be with you; "you are important to me"; and "we can have fun together."
    - Be creative and engage personal interests for non-screen time fun. For example:

        - Engage in sports
        - A bike ride

    - Play Frisbee
    - Cook a meal together
    - Plant a garden
    - Go out for an ice cream treat
    - For very young children teach them nursery rhymes, play *Simon says*, *Peek-a-boo* or *London bridges falling down*
    - Also, ask grand/parents what they did before the Internet for some screen-free fun.

- Invite people to better understand what drives the use of screens.

  - For example, if a teenager checks her Instagram or Snapchat account many times a day, she likely wants to feel connected to her friends. Help in addressing this need to mutually explore other means of feeling connected where she might feel better validated in the absence of those “Likes.”
  - Remember, what works for one person may not work for another. Other examples to possibly consider:

    - A child who is bored and turns to screen usage might benefit from joining a sports team or drama club.
    - The vulnerability a bullied gamer experiences might be met, in part, by

martial arts classes, in addition to speaking with his parents.

- Once the underlying motivation for screen use is better understood, a new routine that satisfies this underlying motive can be sought.
- Efforts to better understand motivations will more effectively address the underlying psychological need.

- Engage children's imaginations for image making

    - Some children say they do not see pictures in their heads of the characters in the book
    - Given that children consume 6-11 hours daily of screen images, they have little practice engaging their own imaginations
    - Consider:

        - CRAFTS—building a model car or boat
        - A school play
        - Making costumes
        - Creative expression by bringing back a family game night such as charades, "Catch phrase", other board or card games

- Service inside and outside the home

  - By contributing to others, children and teens grow up learning their self-worth. We intrinsically wish to give to others. Suggest asking children to contribute by:

    - Feeding the cat,
    - Walking the dog,
    - Emptying the dishwasher
    - Helping select fresh peaches with mom at the grocery store

  - For children to experience service outside the home as meaningful, they must first learn about contribution as relationship
  - It is essential that children feel they belong to us NOT TO the virtual community; this is especially true if this is the community of violent video games

- Consider the challenge to reduce screen time by one hour per week. Then consider a no-screen day. Giving up screens for a day is not intended to be a punishment. It is an opportunity to interact with family and friends in new ways.

  - Choose a day that is realistic to be screen-free.
  - Take some time to reflect on your experience of a screen-free day.

    - What did you notice?

        - How did your no-screen day compare with a day when using screens?
        - What do you notice when you were interacting face-to-face with others for a day?
        - What were you able to do because you were not engaged with your screens?

    - Based upon your experience with a no-screen day, consider finding an entire weekend to go screen-free.

- Need help? For some persons, online gaming or other internet activities become an irresistible obsession.

    - Treatment options range from limited outpatient therapy to intensive residential inpatient programs.
    - Check online for treatment options that fit the need

Section 3

Internet Pornography Addiction: A Catholic Approach to Treatment

Introduction

Over the last two decades, mental health professionals have reported a growing number of people who present themselves for therapy, seeking to address their increased urge to view internet pornography. Accurate statistics regarding the use of internet pornography are difficult to locate; however, available research demonstrates prolific numbers of pornographic internet sites. According to available data, "sex" is the number one search topic on the internet.[76]

Sixty million daily requests are made of pornographic search engines. Forty million adults in the US regularly access Internet pornography with 10% of them admitting to internet pornography addiction.

[76] B. Na, What's the most-searched thing on the internet? Quora. [Accessed May 31, 2019] https://www.quora.com/Whats-the-most-searched-thing-on-the-Internet

Porn hub is a video sharing site which received 2.4 million visitors per hour in 2015 alone. That's 57.6 million per day. This number rose to 80 million daily users in 2017. It is estimated that 64% of college men and 18% of college women engage in online sex every week. Also, 56% of divorce cases involved one party having an obsessive interest in pornographic websites.[77] Researchers now predict that millions of Americans are addicted to internet pornography. In fact, sources assert that the number of users of internet pornography has sky-rocketed, making internet pornography one of the foremost addictions in the world.[78]

Recent surveys report that 90% of teenage boys view online pornography while doing homework.[79] Additionally, young, sexually inexperienced persons, especially males, find it easier to engage in sexual behavior accessing internet pornography than to risk rejection in the context of a face-to-face encounter with a real person.

The expansive reach of Internet pornography has also touched a very young demographic. Notably, the

---

[77] Pornography statistics. Family Safe Media. (2006.) [Accessed May 31, 2019.] http://www.familysafe.com/pornography-statistics/

[78] K. S. Young, X. D. Yue and L. Ying, Prevalence Estimates and Etiologic Models of Internet Addiction. Internet Addiction: A Handbook and Guide to Evaluation and Treatment (John Wiley and Sons, Inc., 2011), pp. 3-6.

[79] J. Ropeloto, Internet Pornography Statistics. Top Ten Reviews. (March 28, 2014.) [Accessed May 31, 2019.] http://www.ministryoftruth.me.uk/wp-content/uploads/2014/03/IFR2013.pdf

average age of children who are first exposed to internet pornography is 11 years old.[80] Exposure to internet pornography in children four to five years old is also being reported. In addition, 28% of boys and 18% of girls have viewed bestiality online and 39% of boys and 23% of girls have seen sexual bondage online.[81] Pornographers use character names such as *Pokémon* and *Action Man* to appeal to children. Once a child clicks on a URL, it opens to a pornography site. When the child clicks again, this action opens to another pornography site. This is what is called mousetrapping.

This section will establish a connection between addictive behavior and pornography. Further, factors contributing to the sky-rocketing prevalence of internet pornography use will be considered. Some neurological effects of internet pornography, which prompt many users to seek more intense erotic images in order to achieve heightened experiences of sexual arousal, will be addressed. Finally, a treatment protocol that can promote affective maturity and reverse the changes in the brain created by viewing internet pornography will be suggested.

Criteria for Internet pornography addiction

Originally, mental health care providers connected the term "addiction" to the use of chemicals such as alcohol, drugs and nicotine. Psychiatrically, addiction

---

[80] Ibid.

[81] Covenant Eyes. [Accessed May 31, 2019.] https://www.covenanteyes.com/

refers to levels of tolerance and withdrawal that hinder affective or psychosocial functioning. These are physiologically mediated symptoms.[82] Tolerance is present when the same amount of substance elicits less response. For example, a person who drinks two glasses of wine is affected by the alcohol. With continued use, the same person will need more than two glasses of wine to produce physiologic reactions. In this way, a person has developed a tolerance for the effects of alcohol. On the other hand, withdrawal refers to the physiologic reaction elicited when the amount of a substance is less present or absent in the body. Withdrawal symptoms include tremors, anxiety, elevated blood pressure and increased rate of pulse and respirations. Affective or emotional disturbances may include depression, irritability, impulsivity, impaired concentration, disrupted sleep or aggressive behavior. These symptoms are observable when an individual is experiencing withdrawal from alcohol or street drugs.

Following a similar pattern of tolerance and withdrawal, experts in addiction disorders describe five successive and interdependent stages through which people pass on the way to an addiction to internet pornography, including discovery, experimentation, habituation, compulsivity and hopelessness. Progression through these stages may be gradual or may occur rapidly after discovering pornographic

[82] Diagnostic and Statistical Manual of Mental Disorders, Fourth Edition (Washington, D.C.: American Psychiatric Association, 1994), pp. 176-9.

websites.[83] In the discovery stage, a person stumbles onto a pornographic website, opening the door for further exploration. Encouraged by the anonymity of electronic transactions, users secretly experiment with sexual material online without getting caught. With repeated exposure—like building a tolerance to alcohol—users develop a habit of sexual fantasies and access pornographic material to increase arousal levels. As users become desensitized to online sex, heightened sexual intensity is necessary to achieve desired levels of arousal. Over time, to avoid life's complications and responsibilities, the habit of accessing internet pornography becomes a compulsion. Sexual excitement becomes associated with tension reduction, relieving feelings of guilt, anxiety or depression. Compulsive internet pornographic behavior is driven largely by tension and agitation, much like an alcoholic is driven to drink at moments of excessive stress. At this stage, men and women jeopardize careers and relationships in order to satisfy their compulsive urges. Despite potential risks, men and women deceive family members and friends to conceal the extent of their involvement with internet pornography, which is no longer a voluntary activity. Restlessness and irritability emerge when attempting to abstain from this behavior. At the height of their compulsion, users of pornography are unable to find the will power necessary to stop and feel hopeless and

---

[83] K. S. Young, Tangled in the web: Understanding cybersex from fantasy to addiction (Washington, D.C.: 1st Books Library, 2001), pp. 40-44.

impotent against persistent and dominant urges to view sexual images online.

Reasons for the addictive nature of Internet pornography

Over the last 40 years, several factors have created a solid platform for the introduction and acceptance of Internet pornography into American society. Principally, the introduction of the birth control pill to the public in the 1960's intensified the heat of the smoldering sexual revolution, as American society separated sexual intimacy from its core purposes of unity and procreation. Artificial contraception reduced sexual intimacy to pleasure-seeking recreation while limiting the possibility of conception. The advent of the World Wide Web made immediate the availability of adult entertainment and amplified the perceived recreational purpose of sexuality. Following this trend, today, more teenagers are becoming sexually active, at a time when they are emotionally vulnerable, uncertain about their moral beliefs and confused about the purpose of sexuality.

Three primary features have added to the appeal of internet pornography. First, the omnipresence of computers, cell phones or other electronic devices allows men and women immediate access to the internet, which makes internet pornography very accessible. Most households today have at least one computer, and most workers have access to a computer at their jobsites, with 20% of men and 17% of women admitting to accessing internet pornography at work. To wit, pornography remains the second most frequent diversion and misuse of the internet in the workplace, after accessing personal email.[84] Second, since many pornographic websites offer free access, they are affordable to viewers of every age and social status. Studies indicate that 80-90% of persons who use online pornography pay nothing, and 10-20% of users pay an average of $60.00 per month.[85] Pornographic preferences generally vary by gender. For example, male viewers seek photographs, videos and live-camera feeds for sexual gratification while women are twice as likely as men to enter chat rooms,[86] seeking friendly conversations that eventually lead to sexually explicit exchanges.

Third, enabling users to pretend to be anyone or no one when accessing pornographic websites, the anonymous quality of internet pornography contributes to the false belief that no untoward repercussions will occur from accessing pornographic sites. The cybersex "relationship" feels more intimate than actual relationships while remaining "safely" anonymous. Fostering dishonesty, fear, self-focus, de-sensitization and self-deceit, men and women addicted to internet pornography value their anonymity and

---

[84] Al Cooper, Irene McLaughlin, Pauline Reich, Jay Kent-Ferraro, "Virtual Sexuality in the Workplace: A Wake-up Call for Clinicians, Employers, and Employees," in Sex and the Internet, 2002, pp.111-112.

[85] K. Doran, "Industry Size, Measurement, and Social Costs", in The Social Cost of Pornography: A Collection of Papers, (Witherspoon Institute, 2010), p. 185.

[86] Porn stats. Covenant Eyes. [Accessed May 31, 2019.] https://www.covenanteyes.com/pornstats/

enhanced sexual stimulation, resulting in increased incidences of compulsive masturbation and more deviant sexual activities, sometimes leading to overt illegal behavior.[87] Accessibility, affordability and anonymity, blending with the change of sexual mores in the use of the internet, contribute to the highly addictive quality of internet pornography.[88]

Neurological implications of internet pornography

Scientific research indicates that alterations in the human nervous system can also result from internet pornography use. The human nervous system[89] has the capacity to build new nerve[90] connections, strengthen-

---

[87] Stephen Arterburn, Addicted to "Love", Understanding Dependencies of the Heart: Romance, Relationships, and Sex (Regal Books, 2003), pp. 117-121.

[88] Al Cooper and Eric Griffen-Shelley, "The Internet: The Next Sexual Revolution," in Sex and the Internet: A Guidebook for Clinicians (Brunner-Routledge, 2002), pp. 5-6.

[89]The human nervous system is made up of the central nervous system (brain and spinal cord) and the peripheral nervous system (nerves in the body that carry messages from the brain and spinal cord to the rest of the body, such as muscles, organs and glands).

[90] The central and peripheral nervous systems are comprised of neurons or nerves. A neuron or nerve has three parts. The portion of the neuron that receives input from other neurons is called the dendrite and is shaped like tree branches. The dendrites lead into the cell body which contains DNA and the other elements necessary to keep the neuron alive. The axon is the cable of the nerve and is of varying length depending on its location and function. The nerves in the leg, for example can be several feet long depending on the height of the person. The neurons in the brain are microscopic in length. These axons are similar to electric cables

ing and increasing some connections while weakening and decreasing others. This change ultimately alters the functioning of the nervous system, modifying the process the brain employs to arrange information. To illustrate, when a child first learns to ride a bike, he initially teeters back and forth on the wheels of his bike until he falls. By repeatedly riding the bike, the child's central nervous system (brain and spinal cord) and peripheral nervous system (nerves in the body carrying messages from the brain and spinal cord to the rest of the body) communicate more effectively so that motor skills necessary to ride the bicycle become more precise and efficient. In this way, the youngster develops skills necessary to balance himself on the bicycle and propel himself forward with a pedaling motion. The brain's capacity to adapt the operation of the nervous system to various stimuli is a normal lifetime phenomenon.

In the same way, studies show that intense and repetitious sexual images initiate modifications at the synapse (the space between nerves through which nerve cells communicate with one another), which changes information processing in the human nervous

---

and carry electric signals toward the dendrites of the neighboring neurons. Axons, or cables of the nerves, do not touch the neighboring dendrites. They are separated by a microscopic space called a synapse. Once an electric signal gets to the end of the axon, it triggers the release of a chemical messenger, called a neurotransmitter into the synapse. The neurotransmitter floats over the dendrite of the adjacent neuron, exciting or inhibiting it. If a neuron receives enough excitatory signals from other neurons, it will fire off a signal. When it receives enough inhibitory signals, it will not fire.

system.[91] I propose that two primary factors are responsible for this modulation.

First, we learned from Marshall McLuhan, a pioneer in the study of the effects of television on the human brain, that more than content or message, the medium of the message plays a significant part in altering the process function of the human brain.[92] In other words, McLuhan's research demonstrates that the volume and rate of delivery of media alter the users' response to stimuli more than the content of the media and radically impacts the response processing function in the user. As such, the effects on the human nervous system from electronic media are increased exponentially with the emergence of high-speed internet. Handheld devices such as iPads and cell phones can alter how the human nervous system processes information because of the speed and volume of electronic stimuli they produce.

Second, neurochemical responses to sexual stimuli factor into the internet pornography addiction equation. Consider the following scenario: Upon viewing, pornography triggers two pleasure centers in the brain of the viewer. Initially, the appetitive or excitatory pleasure system releases dopamine, which the body experiences with enjoyable activities, such as laughing, eating a good meal, running or being sex-

---

[91] M.J. Koepp, R.N. Gunn, A.D. Lawrence, V.J. Cunningham, A. Dagher, T. Jones, D. J. Brooks, C.J. Bench, and P. M. Grasby, Nature, 1998, vol 393, no. 6682, pp 266-268; Doidge, Norman, The brain that changes itself (New York: Penguin Books, 2007), p. 309.

[92] W.T. Gordon, ed. Understanding media: The extensions of man, Critical Edition (New York: Ginkgo Press, 1994), p. 19.

ually aroused. Acetylcholine[93] is also discharged into the brain, which plays an important role both in learning and memory and helps the brain focus and form sharp recall of pleasurable experiences. Acetylcholine stores these pleasurable images in the brain and makes them readily available for recall. Secondarily, the consummatory pleasure system, which produces sensations of calm and satisfaction after recreational experiences, secretes oxytocin and serotonin, bringing feelings of serenity and bonding. Endorphins are also released heightening euphoria.

Beyond viewing pornography in magazines or film strips, viewing pornography over high-speed internet, with its capacity for delivering rapid bursts of endless images, hyper-activates the appetitive pleasure center, powering a surge of dopamine through the pleasure center and forcing the attentional and motivational mechanisms of acetylcholine to focus most intensely on what is triggering the dopamine surge. With repeated excessive stimulation, the brain adapts to the stimuli dampening the pleasure system's ability to respond not only to the same sexual stimuli but to all ordinary pleasures.

---

[93] Acetylcholine plays a central role in the health of the brain. It is stored in the nerve and can be released into the synapse once the nerve is activated. Acetylcholine has several functions. For example, the parts of the brain involved in memory, learning and mood use acetylcholine extensively. It is also responsible for sending messages from the brain to certain muscles causing them to move, helps maintain optimum muscle tone, ensures that mucous membranes are always lubricated and moist and helps to schedule REM or dream sleep necessary for restful sleep.

Under-stimulated, the addict needs more and more novel internet pornography to experience pleasure. For example, pornography addicts may turn to child pornography, not because they are clinical pedophiles but because they have become habituated and desensitized to other forms of pornography.[94] Introducing new pornographic images, such as scenarios of sex with violence or humiliation, sparks the release of more dopamine and generates more arousal. They keep watching because they like the "pleasure rush" of dopamine release, dismissing the significance of how the changes in the pleasure centers of their brain have altered what sexually arouses them.

Acetylcholine, which both stores pleasurable images in the brain and makes them readily available for recall, along with oxytocin, a neurochemical that contributes to creating an enduring experience of pleasure, combine to create an associative quality to internet pornography use. To illustrate, a pianist who becomes sexually aroused when viewing internet pornography while playing classical music can associate classical music with sexual stimulation. Additionally, typing on a computer keyboard can remind him of playing the piano, which also triggers sexual arousal. At this juncture, the addict can "play the keyboard" of his computer to become sexually aroused.

---

[94] P. Paul, Pornified (New York: Owl Books, 2005), pp. 196-198; Quale, Ethel, and Max Taylor, Child pornography and the internet: Perpetuating a cycle of abuse." Deviant Behavior vol. 23, no. 4 (2002), pp. 331-361.

Addictions often begin with voluntary choices to engage repeatedly in a behavior, like the choice to consume alcoholic beverages, smoke cigarettes or use drugs. Over time, the body builds a tolerance to the stimuli and thresholds of consumption increase. In the same way, internet pornography use may begin with occasional choices to view sexual images. However, high-speed internet and the release of neuro-transmitters, like dopamine and acetylcholine, blend to create a powerful force that temporarily pleases while creating a malfunction in the pleasure centers. The addict's neurons, assaulted by abnormally high levels of dopamine, respond defensively by reducing the number of receptors to which dopamine can bind. The addict experiences a dopamine deficit. While internet pornography addicts begin using internet pornography for pleasure, they end up viewing pornography to alleviate feelings of depression and irritability. Reclaiming power and control over the pleasure centers of the brain will require a formidable effort.

A case study

Thomas is a 19-year-old college freshman who presented himself for psychiatric evaluation and treatment referred by his family physician. He was dating a woman he believed "was the one God wanted me to marry." He was worried, however, that if she ever found out he was addicted to internet pornography, she would "dump me." He had made an appointment to speak with his doctor but was too embarrassed to talk openly about his concerns. His doctor thought Thomas suffered from depression and anxiety. He valued his faith and was seeing a spiritual director monthly.

Thomas first looked at internet pornography at age 9 "out of curiosity." He was aroused by the images. The experience was pleasurable and embarrassing. He began to search for more intense stimulation by exploring various pornography sites. Thomas told his spiritual director that he was struggling with internet pornography use. His spiritual director instructed Thomas to pray and to stop viewing pornography. However, he was not given specific direction on how to address the problem nor did his spiritual director ask him more questions about his reported concern.

After countless relapses, Thomas gave up efforts to resist pornography use. He avoided being honest with anyone about his concerns because he felt like a failure. He believed that he was helpless to address his problems and became hopeless about being free of his addiction.

When Thomas presented himself for evaluation and treatment, he said this was his "last resort." He had been in therapy before to address issues of self-esteem. He was unconvinced that therapy could really help him. He privately agreed to therapy so that he could say one more time that therapy did not help. He said, "I have tried everything, and I am sick of trying." He also acknowledged the conflict between his use of pornography and his desire to be a faithful future Catholic husband to his Catholic girlfriend, but Thomas enjoyed the sexual pleasure. He was ashamed of the behaviors that resulted in his fantasizing about women he had viewed on the internet. His guilt and shame kept him at a distance from his girlfriend, and he was even avoiding some of his family members. He was preoccupied with sexual images. This also began to affect his capacity to concentrate and his interest in pursuing an engineering degree. He was missing college classes. He told his girlfriend that he had a sleep problem and depression, which his doctor was trying to treat.

At the beginning of treatment, Thomas admitted he had mixed feelings about becoming free of his internet pornography addiction. He reported, "Most all of my life I have used pornography to survive, deal with my frustrations and loneliness. The thought of never looking at pornography again scares me." We spoke of the fact that strategies for healing from internet pornography addiction were not complicated. The key, however, was whether the person addicted was willing to do the work of facing the hurts, shame, guilt and anger of his past and to be open to the grace of God.

In order to begin the healing process, Thomas had to surrender control; he was still trying to heal himself. Surrendering control was not giving up but giving himself over to the graces of God in the sacraments, rebuilding his relationship with his family, developing his relationship with his girlfriend and establishing a therapeutic process. To begin, he had to be honest with himself and God about his problem, and he had to be willing to allow at least one other person to walk with him into what he called his "black box" and to keep walking toward the Light.

Over the next several weeks of intensive therapy, we reviewed in detail his patterns of internet pornography use and his early childhood history. Thomas also gradually committed to being honest with his family, spending time with his Catholic friends, reading Sacred Scripture, engaging spiritual reading, going to regular confession and spiritual direction, even during many distractions and temptations to act out sexually.

We discussed Thomas' triggers and the false beliefs that were keeping him shackled in his pornography addiction. This is commonly called the Reaction Sequence[95]:

- Trigger (stimulus)
    - He sees a scantily clad attractive woman on the cover of a magazine in a grocery store aisle while checking out groceries or the "dancing" images of women as he opens his email account
    - Once the stimulus is triggered, his brain gives meaning to the stimulus which elicits a reaction of sexual arousal

---

[95] K.B. Skinner, Treating pornography addiction: Essential tools for recovery (Orem, UT: GrowthClimate, Inc., 2005), p. 45-55.

- Emotion
    - He becomes more excited
    - He experiences a growing urge to search for titillating images
- Imagination
    - Sexually arousing images begin to fill his mind
    - This triggers the thought, "I like what I am feeling"
- Chemical release
    - Neurochemicals begin to rush through his body
    - NOTE—once a reaction sequence is fully developed in the mind, these chemicals are released into his body before he ever sees pornography
- Physiological changes
    - With the release of these chemicals, his heart rate increases
    - His hands become sweaty or cold
    - His eyes dilate
    - His muscles become tight

- Battle of the will
    - Lust contributes to ambivalence
    - That is, a twofold will
- I shouldn't look at pornography
- I have done this many times; one more will not matter/no one will find out
    - At this point, Thomas has waited too long to engage in the battle against lust
- Sexual pleasure becomes an end in itself
- Lust plunges him into sexual sins
- Lust has blinded his reason, and he has become indifferent to the interior struggle
- The narcissism or self-focus of lust creates in him an aversion for the things of God
- His mental wanderings have given free rein to his imagination
- He no longer exercises self-control. He succumbs to self-soothing sexual behaviors authorizing the following beliefs:
    - "I cannot get over this problem so why try"; "It isn't that bad"; "One more time won't hurt"
    - He has accepted his behavior as something that is "okay"

- Response
    - 12-14 pornographic sites are clicked onto his computer screen through which he navigates for more exciting images
    - Videos are next with their sights and sounds
    - Masturbation intensifies his sexual stimulation, and he "loses" himself in the sensual pleasure

Through this process of detailing his reaction sequence, Thomas gained insights into his addiction and the power that he had given to pornographic images.

There is no quick or easy way to erase from the memory pornographic images associated with cybersex activities. Spiritual, mental, emotional and physical discipline is necessary to be free from these images and behaviors. Thomas began to realize that to stop viewing pornography requires a keen self-awareness. He also learned that pornography is not something that is simply overcome through exerting strong willpower. He was reminded that overcoming his addiction was a process and not an event.

The second component of the healing process came when Thomas developed a strategy to deactivate his reaction sequence[96].

---

[96] Ibid., p. 51-55.

1. He was asked to write down the reaction sequence that led to the behaviors he wanted to change. This included:

    - Mapping out his personal reaction sequence
    - Naming for himself triggers, emotions, images, thoughts and physiologic reactions as well as beliefs he held about his behaviors
    - Writing down times, places, situations and company kept during which he was most vulnerable to giving in to pornography

2. He next wrote down the specific behaviors he planned to change in each part of his reaction sequence. This included:

    - Taking a different route home
    - Breaking routines that have resulted in his accessing internet pornography. For example:

        - Channel or internet surfing
        - Telling himself he deserved a break when feeling overwhelmed, frustrated, lonely, or unappreciated in his relationships
        - Idle time
        - Procrastination
        - Cruising by store magazine racks in local shops, entering video rental stores

- When one of his triggers for seeking pornographic images presented itself, he planned alternative activities such as:

  - Focusing on attending to a need of his family
  - Calling a friend
  - Praying for God's assistance
  - Doing some physical work
  - Imagining how he felt after having viewed pornography. For example, imagining his life without his family and friends and considered how to express to them how much he values them

3. Next, he was asked to commit to reviewing his plan of action each time his reaction sequence was set off to assess the effectiveness of his action plan
4. This process contributed to a newfound capacity to deactivate his reaction sequence.
5. He also identified beliefs that give him "permission" to view pornography.

Identifying his reaction sequence and writing down his plan of action empowered Thomas to break down his addictive behaviors and avoid or deactivate his trigger points.

The third component of the healing process came when Thomas began to explore the power that he had given his distorted or negative beliefs. Throughout the years, he had formed negative or distorted beliefs about himself which evolved from childhood experiences. These beliefs fueled his relapses. Some of his key beliefs were:

- I am inferior, inadequate, unlovable
- I am alone
- Nobody understands how hard I am trying, but I cannot stop
- I have repeatedly told myself that I would not relapse again, but I keep going back to it
- I deserve some pleasure that does not hurt anyone else
- I am a bad person
- Nobody would like me if they knew who I really was
- I cannot meet the expectations of a Catholic and good husband

These beliefs kept Thomas trapped in what he referred to as "my black box." He had not realized this until he wrote down these negative or distorted beliefs and saw how these thoughts were hopelessly guiding his daily living. We worked together to have him challenge each negative belief about himself and work at bringing reality to bear on his problem.

Today, Thomas continues to take radical steps in order to avoid both the remote and the approximate occasions of sin. He makes daily a heartfelt decision to break free of internet pornography and then acts on that decision. He has purged his computer of all pornographic images, moved his computer to an open location, uses internet minimally, installed a filter, has an accountability partner whom he calls every day as well as when he is tempted to view pornography. His anchor is now his prayer life and sacramental life. He has committed to daily sharing with his friends and family, reading Sacred Scripture, spiritual reading, regular confession and spiritual direction. He strives to make acts of faith, hope and charity and to grow in the little virtues. He has begun to reclaim his self-esteem as a Catholic college student, but he still faces a daily battle with pornography. An essential factor to his continued human growth and development is that his struggle is out in the open now and he is working, one day at a time, to resist the urge to view pornographic images.

Conclusion

With inexhaustible abilities, the internet can be a powerful tool for learning and communicating. In his address on the 2002 World Day of Communications, Pope John Paul II wrote:

> The Internet causes billions of images to appear on millions of computer monitors around the planet. From this galaxy of sight and sound will the face of Christ emerge and the voice of Christ be heard? For it is only when his face is seen and his voice heard that the world will know the glad tidings of our redemption. This is the purpose of evangelization. And this is what will make the Internet

> a genuinely human space, for if there is no room for Christ, there is not room for man. Therefore...I dare to summon the whole Church bravely to cross this new threshold, to put out into the deep of the Net, so that now as in the past the great engagement of the Gospel and culture may show to the world 'the glory of God on the face of Christ' (2 Cor 4:6). May the Lord bless all those who work for this aim. (John Paul II, *Internet: A New Forum for Proclaiming the Gospel,* #6).

The new frontier of the second millennium, cyberspace is replete with interplay of danger and promise. Armed with the vision of our transcendent dignity as persons, each one of us is summoned to the great adventure of using the internet's potential to proclaim the Gospel of Christ.

Section 4

What Parents Ought to Know About the Hottest Social Media Apps Teens are Using[97]

This section describes some of the most popular digital applications and websites used by teenagers. Parents do not need to know all about each application or website to provide a safety-net for their children. Knowing the basics such as what an app does, why it is so popular and what problems can occur if it is not used responsibly will help parents communicate with their teenagers on how to make choices that will be a positive force in their lives rather than a negative experience for them, yet without smothering them.

Kik Messenger was founded in 2009 by a group of students from the University of Waterloo in Canada and was released in 2010. Kids text for free on this app. It is fast and has no message limits. Because it is an app, the texts will not show up on a teen's phone

[97] Review for what your kids want to watch (before they watch it). Common Sense Media. [Accessed May 31, 2019]. https://www.commonsensemedia.org

message service and parents will not be charged for this messaging service beyond the standard data rates.

What parents need to know:

- It allows multiple people to submit pictures or screen shots
- It can display a user's full name
- It is popular among teens for flirting and sexting
- Since it is impossible to verify someone's identity on *Kik*, it has the potential to attract online predators

ooVoo was founded in 2006 by Ohio entrepreneur Clay Mathile. *OoVoo* is a free video, voice and messaging app. Users can have group chats with up to 12 people, free of charge. It is common for teens to log on after school and to keep this app open while doing homework.

What parents need to know:

- Users can only communicate with those on their approved contact lists which can help ease parents' safety concerns
- It can be distracting because the service makes video-chatting very affordable and accessible
- It also can be addicting

Instagram was created in October of 2010, By Kevin Systrom and Mike Krieger, both from Stanford University. It has over 200 million users who post over

60 million photos daily. This app has dozens of filtering systems with many editing options which aide users to produce the "near perfect" pictures vying for the most "Likes" for their picture.

Instagram allows users to:

- Snap, edit and share photos and 15 sec videos
- Apply fun filters and effects to their photos

Photos and videos on *Instagram* are public unless privacy settings are engaged. Hashtags and location information make photos visible beyond a teen's followers if the account is public. *Instagram* now allows "private messaging" for up to 15 friends.

Teens are often on the lookout for "Likes". The online persona is needier than the real one. If they don't get enough "Likes" for their selfie as expected, they decide to post another, but if they still do not receive a good response this can affect their self-confidence. Society, after *Instagram*, has become infatuated with taking pictures, uploading them onto the app, changing what they see with the filters and editing possibilities, suggesting that *Instagram* has created a subconscious notion that people and places are not good enough in their natural form. Teens become dissatisfied with themselves and their bodies. Consider this example:

Essena O'Neill[98], despite having more than half a million followers on *Instagram*, 200,000 on *YouTube* and *Tumblr* and 60,000 on *Snapchat* announced she was quitting social media for good. The then 18-year-old Australian who was living what seemed like a "perfect" life wrote on an *Instagram* post: "I'm quitting *Instagram, YouTube* and *Tumblr.* I deleted over 2000 photos here today that served no real purpose other than self-promotion. Without realizing it, I've spent the majority of my teenage life being addicted to social media, social approval, social status and my physical appearance...Social media, especially how I used it, isn't real. It's contrived images and edited clips ranked against each other." In her *YouTube* video, O'Neill spoke about how unhappy her social media obsession made her. "I spent hours watching perfect girls online, wishing I was them. Then when I was 'one of them' I still wasn't happy or at peace with myself". She renamed her last *YouTube* video post "Social Media Is Not Real Life." She chose to edit the captions on photos she kept in order to reflect the truth of what happened "behind" the image. Now, this teenager is launching her own website called *Let's Be Game Changers* where she hopes to continue to educate people about the destructive nature of trying to gain approval online. "I know you didn't come into this world just wanting to fit in and get by. You are reading this now because you are a game changer, you

[98] M. McCluskey, Teen Instagram star speaks out about the ugly truth behind social media. Time. (November 2, 2015.) [Accessed May 31, 2019.] http://time.com/4096988/teen-instagram-star-essena-oneill-quitting-social-media/

might not know your power yet, I am just finding mine, but man… when you do… far out you'll go crazy. It'll be brilliant. You'll be brilliant," she wrote.

Snapchat, launched in 2011, was developed by Evan Spiegal and two other Stanford University students. This was Spiegel's final project for his design class. He initially named this application "Peekaboo" photo messaging because it produced what he called "impermanent photos." Users take photos, record videos, add texts and drawings and send them to a controlled list of recipients. The sender sets a time limit for how long the recipient can view their photo messages, which can range from 1-10 seconds before they are deleted from the server. Spiegal renamed the application Snapchat. By May, 2012, twenty-five images per second could be sent. By November, 2012, one billion photos had been shared with twenty million being shared per day.

Snapchat lets users put a time limit on the pictures and videos they send before they disappear. Many teens use the app to share goofy or embarrassing photos. It is a myth that *Snapchats* go away forever. The seemingly "risk-free" messaging might encourage users to share inappropriate (often sexual) images. Also, the total number of sent and received chats can be added up for viewing. Friends can see scores of who has most views.

Tumblr is an app that streams text, photos and/or videos and audio clips. It is easy to find pornographic images and videos, depictions of violence, self-harm, drug use and offensive language in this app. Users must first create a profile that can be seen by anyone online. For privacy, a second profile has to be created. Posts are often reblogged. Do parents really want their child's words and photos on someone else's page?

Vine allows the user to post and watch looping 6-second video clips. In a few minutes of random searching, the user will most likely come across nude pictures of men and women as well as persons blowing marijuana smoke at each other. Teens have created videos of family members and posted them. Who knows how they were modified?

YouNow allows users to live broadcast from anywhere. Users can search topics like "dancing" or "singing". *YouNow* performers are often teenagers. Viewers can chat with performers. Viewers can tip with real money if you "Like" the performer. Performances are often of a sexual nature. Teens live stream themselves sleeping and viewers are watching by the thousands.

Whisper is an anonymous social network. It has been called an "online confessional" app because it allows users to post whatever is on their minds paired with images without sharing any identifying information. Most content is sexual, about depression, substance abuse, lies told to employers/teachers or personal feelings of insecurity. A "meet up" section encourages sharing of personal information.

Yik Yak is a free social networking app. Users post comments to the 500 geographically nearest *YikYak* users. No identifying information other than a user's location is required. Posts frequently are sexual, about drug or alcohol use and can contain cruel or toxic comments or cyberbullying. Posts are voted on by other users. Two or more down votes removes the post. School lock downs have resulted due to posted threats.

Omegle is all about putting two strangers together to chat via text or video. No registration is required for this app or site. "Interest boxes" allow users to select a potential chat stranger partner by shared interests. Filled with persons searching for sexual chat or video, *Omegle* offers links to porn sites. The anonymity of this app makes it attractive to some teens and potentially harbors sexual predators.

AfterSchool is on over 23,000 high school campuses. It is only accessible to teenagers and requires verification that the user is in high school. The high school where a post is generated is the only identifying information given. *AfterSchool* does not identity who wrote the post. The anonymity fosters cyberbullying, sexting and alcohol, tobacco or other drug chatting.

Ask.fm is a question and answer site that thrives on anonymity. Participants create profiles (real or not) that anyone, not just site members, can ask questions about. *Ask.fm* has over 120 million users, fifteen million of whom live in US. The website is based in

Ireland with satellites in Riga, Latvia and Oakland, CA. The founders of the website are Mark and Iga Terebin who are two sons of a wealthy Red Army officer. This site is integrated with *Facebook* and *Twitter*. What is shared on *Ask.fm* is therefore easily shared on other sites. Questions asked are frequently sexual and or humiliating and can be screen shot and sent to friends for embarrassment. This site is "associated with some of the worst forms of cyberbullying and has been linked to numerous suicides around the world".[99]

Voxer was created in May 2011 by Tom Katis and Matt Ranney. Katis was a special forces communication sergeant in Afghanistan. After his tour of duty in 2007, he designed *Rebelvox* for better communications on the battlefield between soldiers which turned a smartphone into a walkie talkie. In addition to sending voice messages back and forth, *Voxer* users can also send texts, photos, or location information. This messaging app has been used for cyberbullying.

Blendr is an anonymous flirting app used to meet new people through GPS location services. People can send messages, photos and videos, and they can rate the attractiveness of other users. There are no

[99] R. Cooper, 'Bullying is relentless these days – there is no break from it': Mother's campaign to close Ask.FM after her daughter's suicide because of online abuse. DailyMail. (November 19, 2013.) [Accessed May 31, 2019.]
https://www.dailymail.co.uk/news/article-2509874/Mothers-campaign-close-Ask-fm-daughter-Izzy-Dixs-suicide.html

authentication requirements. Blendr is often used for cyberbullying by a group of kids targeting another child to purposefully make his or her rating go down. Sexual predators can also contact minors and minors can meet up with adults via this app.

Hot of Not is a rating app that lets viewers judge the attractiveness of others' pictures signaling a heart sign or an X.

#tbh appraises others' selfies. For example, "Am I pretty?" or "I think you're really pretty".

"To Be Honest" is a term that encourages online users to express honestly how they feel about a person or an idea they post. For example, someone might post a photo or thought and others might respond with: "TBH, you are really pretty even though we don't talk as much as I'd like to" or "I've never told you this but TBH, I think the way you play guitar and write music is amazing." TBH carries both the power to lift another person up or to put another down. TBH has become so popular that entrepreneurs have introduced a TBH App *Facebook* Page with instructions on how to write a TBH to "find out what your friends REALLY think about you."

Additional trending terms which can be helpful to know about:

JBH = Just Being Honest
LBH = Let's Be Honest or Loser Back Home
TBBH = To Be Brutally Honest

SMEXI = Smart and Sexy
IMO = In My Opinion
GOMB = Get Off My Back
KOTL = Kiss On The Lips
KOS = Kill On Sight. This is a term that originated with online war games such as *World of Warcraft*. It means basically marked for death just by showing your face. However, it can also be used as a threat by a cyber bully.
S&D = Search and Destroy (also could be a threat)
Ug = Ugly
CID = Acid (as in, the drug)
WAW = What a Waste
CNBU = Can Not Be Unseen
Gomer = Geek, wierdo, nerd
Ratchet = Ugly, nasty, awful
Broken = Hungover from alcohol
Beep face = A general insult
Butter face = Refers to a person with an alleged pretty body but ugly face
420 = This means marijuana (also look for words like 420 4life, boo, blunt, and Buddha)
ASLP – Age, Sex, Location, Picture
FYEO = For Your Eyes Only
CD9 = Code 9; parents around
POS = Parents Over Shoulder
Sugarpic = Suggestive or erotic photo
53x = Sex

Recommendations:

Parents can help guide their children in using apps for fun and connection and not fuel for self-doubt. Take an interest in your children's online activities. Engage in conversation with your children's gaming and app practices, successes and struggles.

Don't assume your children understand the digital risks they might encounter in these apps. Children may have tech skills but lack the wisdom needed to navigate risky digital interactions. Eventually, most children will find themselves in the middle of a sensitive situation. Help them to develop a healthy awareness of potential problems, not be afraid to speak with you about their concerns and learn virtue based social etiquette online in situations they might encounter.

Here are a few suggestions:

- Be a healthy role model
- Guide children in posting constructive comments to their friends
- Actively enter their world by playing games/using apps with your children
- Get to know the social media sites that your children are using. Explore the site, and educate yourself on what each offers and how children are using it
- Talk about the pictures they post
- Ask how feedback makes them feel
- Help them develop a healthy self-image

- Set limits
- Place computers in a public room
- Discuss sexting
- Talk about cyberbullying
- Report abusive behaviors
- Warn them about strangers
- Maintain strong passwords
- Keep your home network updated, clean and protected

Section 5

Is Electronic Media Making Us Smarter?

Before the advent of the internet, we were never able to jump between one link of material to another. Today, we begin a search and another search engine draws our attention to a few words or sentences related to our search. We click on the link associated to these words and the focus of what we are searching for is expanded. Relevant information becomes harder to distinguish from less important or irrelevant information. The influx of messages that we receive online begins to overload our working or short-term memory. It is like trying to read a book while doing a crossword puzzle. The process of memory consolidation is hindered. Organizing information brought on the screen by various search engines and storing it into patterns of knowledge is decreased and less is integrated as a concept or principle to be stored in our long-term memory.

Jakob Nielsen, a web design consultant who conducted an eye tracking study of *World Wide Web* users, found that a vast majority of users skimmed a text quickly in an F pattern.[100] That is, they glanced all the way across one or two lines of a text, then dropped down to the next lines of written material which are scanned across only half way. This study estimated that a person reads 18% of a very wordy page on the *World Wide Web*, thereby likely missing salient points contained in the script. Such an approach to reading on the *Web* does not increase knowledge, integration of material nor comprehension.

Certain cognitive skills, however, are strengthened by the *World Wide Web*. These tend to involve more primitive mental functions such as eye-hand coordination, reflex responses and the processing of visual cues.

What about wireless laptops in the classroom?

Researchers are now looking at the effects of computer use on classroom learning. Many college professors are finding that they must choose whether or not to restrict the use of laptops and other electronic devices in their classrooms or let their students decide this for themselves–something they never thought about ten to fifteen years ago.

Most often, it is the students who decide between taking notes electronically or by hand. It is a given that many students who have an electronic device in the classroom do not use it solely for taking class notes, but

---

[100] J. Nielsen, F-shaped pattern for reading web content (original study). NN/g Neilsen Norman Group. (April 17, 2006.) [Accessed May 31, 2019.] https://www.nngroup.com/articles/f-shaped-pattern-reading-web-content-discovered/

also to check their *Instagram, Facebook* or *Twitter* accounts or surf the *Web* to check the news or find out what is new on *eBay*, for example. Studies demonstrate that such multitasking reduces student learning.[101] This is so not only for those students using their computers or tablets, but also for their neighbors who become distracted by the flashing of lights from screens surrounding them.[102]

How do we address all of this, given the fact that most of our schools today teach with technology? What are studies telling us about the effects of computer use on learning? Some studies looked at the correlation between the device used (computer, iPad or pen) and the learning that took place. Some studies assigned the use of computers to one group of students and another group to take notes by hand. The problem with the first approach is that all students do not use a computer to learn in the same way. For example, it could be that the students who use computers or tablets to take notes are more easily bored or more easily distracted and therefore they exhibit lower performance on exams than their peers who choose to take notes by hand.

In this scenario, the lower academic performance is not directly related to the use of a computer to take notes in the classroom. In addition, the flaw with the comparative approach of students assigned to use a

---

[101] H. Hembrooke, The laptop and the lecture: The effects of multitasking in learning environments, Journal of Computing in Higher Education Volume 15, Number 1, 2003 ISSN 1042-1726, p. 46-64.

[102] Stop Multitasking! It's Distracting Me (And You), August 19, 2013 10:49 AM ET Commentary Tania Lombrozo, NPR.

computer and those assigned to take notes by hand is that such experiments generally study the students only over a few hours.

A more accurate study of the effects of computer use on student learning would require studying a large group of students randomly assigned to identical courses that vary only in whether the students are allowed to use computers or tablets during class. West Point[103] did such a study looking at 726 sophomores in an introductory economics course. The students were randomly assigned to three versions of the same course differing only in the manner students were permitted to take class notes: No devices allowed in the classroom, use electronic devices all you want, or must get permission to use a tablet or computer "face up" on the desk. This study continued through the duration of the semester course. All students took the same exams, both short answer and multiple-choice questions. The exams were graded by an automated system for objective measures of learning assessed across all students of all three versions of the course. The results showed that those students who were in the course version that allowed the use computers or tablets performed significantly worse than their peers who took the course version where computers were not permitted. The researchers further considered composite ACT and baseline GPA scores. There were no statistically significant score differences in the three

[103] Susan Payne Carter Kyle Greenberg Michael Walker, Working Paper #2016.02 The Impact of Computer Usage on Academic Performance: Evidence from a Randomized Trial at the United States Military Academy, May 2016.

groups of students. Based on this finding, West Point concluded that "removing laptops and tablets from a classroom is equivalent to improving the quality of the teacher by more than a standard deviation".[104]

This United States Military Academy study and others like it raise the question: What is it about the use of computers in the classroom that decreases student learning? The authors suggest a couple of possibilities. First, students are using the computers and tablets for tasks unrelated to the course, such as accessing social networking sites or emailing or surfing the *Web*. Second, taking notes on a computer is less effective for learning material than handwriting class notes.[105] Students who take notes on a computer tend to type verbatim what they hear without engaging the deeper processing required for conceptual learning. At exam time, students who took notes by hand did better than those who used computers because of the necessary deeper learning engagement required of handwriting notes as compared with typing notes. Third, the authors speculated that teachers and students interact differently in a classroom where students are most often looking at their computer screens.

Researchers in Norway looked at multitasking not on computers but on cell phones, and found that doing so created generational boundaries which undermined

---

[104] Ibid.

[105] Pam Mueller and Daniel Oppenheimer, The Pen Is Mightier Than the Keyboard Advantages of Longhand Over Laptop Note Taking, Journal of Experimental Psychology Learning, Memory, and Cognition, 36, April 23, 2014, 1118–1133.

family rituals and shared communication, magnifying the importance of the peer group while decreasing the importance of family.[106] Elinor Ochs, anthropologist at UCLA's Sloan Center, also found that multitasking contributes to decreased family interaction.[107] These social costs may be contributing to the sense of loneliness described in iGens due to a decline in their sense of belonging.[108]

Suggestions

What can minimize some the cognitive and social effects of multitasking? One answer is reading. Research has shown that the amount of out-of-class reading done in college years is a statistically significant predictor of critical thinking skills. In addition, reading promotes imagination, increases vocabulary and encourages reflection. Also, engage-ment in more face-to-face activities helps to promote bonding to those with whom we live and those whom we encounter in our daily lives.

---

[106] P. Greenfield and Y. T. Uhls, Explore research on how media multitasking affects productivity and development. Common Sense Media. (January 21, 2010.) [Accessed May 31, 2019]. https://www.commonsense.org/education/blog/expert-article-kids-and-multitasking

[107] C. Mulligan, The truth about teens and multi-tasking. The Children's Digital Media Center. (November 23, 2011.) [Accessed May 31, 2019.] https://cybersolutionstoday.blogspot.com/2011/11/truth-about-teens-and-multi-tasking.html?m=0

[108]

Perhaps most importantly, encourage breaks from or avoid multitasking. For parents, this may mean setting limits on media time or turning off the TV and pulling out a board game that requires concentration on a single task. But, more generally, it means slowing down the pace a bit and encouraging family time since the positive influences of parents on a child's development is most significant when parents spend time with their children.

What could our schools do?

It is important to ask what are the situations in which using computers would be a greater benefit. Technology is being incorporated more and more into educational settings to enhance learning. The focus should not be to teach with technology but to use technology to convey content more powerfully and efficiently. Well thought out, enhanced educational experiences can be the outcome. This can certainly add a burden to many teachers' classroom preparations.

For children, consider making part of the classroom assignment to collect *YouTube*, audio files, websites or other online material related to the topic of instruction. For example, if a teacher is teaching a unit on the history of Saint Louis, Philadelphia or British Colombia, students could be divided into groups to look up links on photo collections, podcasts, YouTube videos and other multimedia presentations on events that occurred during a certain historical time frame in the geographical area being studied. A teacher could then assign the students to present their group work to the rest of the class. Such assignments that require integration of material for deeper comprehension and teaching others about what has been learned help to create richer educational experiences for students.

We need to take advantage of this new generation of children and young persons who love technology to refocus how these persons are educated. In doing so, students are more likely to engage in what they are learning. Teachers can access an enormous amount of curriculum content online in a variety of formats, including audio and video pieces that can help bring the material to life for students. These materials are often free.

Helpful sites include:

- DiscoveryEducation's Lesson Plan Library ((http://school.discoveryeducation.com/lesson plans)
- Teachers Helping Teachers (http://www.pacificnet.net/~mandel/index.html)
- TeachersFirst.com (http://www.teachersfirst.com/index.cfm)
- Thinkfinity (http://www.thinkfinity.org/lesson-plans)
- Institute for Theological Encounter with Science and Technology (http://www.faithscience.org/creationlens)

Section 6

Engaging the Human Genius to Create Apps that Promote Neurocognitive Health (rather than create more apps designed to become addictive)

Although some neuroscience has placed its focus on developing apps and gaming devices that are addictive, certain groups of researchers today are focusing their efforts at harnessing the captivating characteristics of technology to foster healthy human development. The MIND Institute (Medical Investigation of Neurodevelopmental Disorders) at UC Davis is one such facility. Impulsivity is characteristically seen in Attention Deficit Hyperactivity Disorder (ADHD) and other neurodevelopmental disorders. The MIND Institute is exploring whether scientifically-designed games could help attenuate some of the deficits of ADHD and similar disorders.[109]

---

[109] A. Swillen, E. Moss, S. Duijff, Neurodevelopmental outcome in 22q11.2 deletion syndrome and management, AJ Med Genet A, 2018 Oct;176(10):2160-2166. doi: 10.1002/ajmg.a.38709.

Dr. Simon, a neuroscientist at UC Davis, is developing video games specifically aimed at enhancing a player's capacity for spatiotemporal cognitive abilities. Dr. Simon makes the important distinction that playing commercially available action games can build skills, but such games do not necessarily build capacity: "If I flip a coin a hundred times, I can get good at flipping coins, but it doesn't help me play the piano." Dr. Simon and his coworkers have developed spatiotemporal cognitive capacity building video games that act as "digital medicine" to treat children and adults with spatiotemporal cognitive impairments.[110] The current treatment for school-aged children with ADHD is the prescription of stimulants and other medications whose side effects can include problems with sleep and appetite. Although much research needs to be done, the early pilot testing of this "digital medicine" looks hopeful.

The MIND Institute is also studying digital technology to better understand distractibility in developmental disorders and some mental illnesses. Their aim is to develop targeted treatments for inattention and hyperactivity which often have devastating ramifications on social and work performance.[111] Other researchers are exploring the

---

[110] A. Fell, Startup Cognivive plans games as digital therapies. UC Davis. (November 6, 2017.) [Accessed May 31, 2019.] https://www.ucdavis.edu/news/startup-cognivive-plans-games-digital-therapies/

[111] S. Benyakorn, C. A. Calub, S. J. Riley, A. Schneider, A. M. Iosif, M. Solomon, D. Hessl, J. B. Schweitzer, Computerized cognitive training in children with autism and intellectual

potential benefit from video game therapy in brain injury victims who have difficulty controlling motor skills and for persons with memory deficits associated with aging processes.[112]

There is much that is still unknown about how daily experiences and digital devices are affecting human development socially, emotionally, intellect-tually and physically. To wit, the National Institute of Health has launched the largest government study to date on adolescent brain development referred to as the Adolescent Brain Cognitive Development (ABCD) Study. It will follow 11,874 youth, ages 9-10, including 2,100 of whom are twins or triplets throughout their young adult life. This landmark study on brain development and child health will use advanced neuroimaging to observe brain development in children throughout adolescence, while tracking social, behavioral, physical and environmental factors that may affect brain development and other health outcomes. The study is being conducted at twenty-one

---

disabilities: Feasibility and Satisfaction Study", JMIR Ment Health, 2018 May 25;5(2):e40. doi: 10.2196/mental.9564.

[112] S. Straudi, G. Severini, A. Sabbagh Charabati, C. Pavarelli, G. Gamberini, A. Scotti, N. Basaglia, The effects of video game therapy on balance and attention in chronic ambulatory traumatic brain injury: an exploratory study, BMC Neurol. 2017 May 10;17(1):86. doi: 10.1186/s12883-017-0871-9; J. P. Cuthbert, K. Staniszewski, K. Hays, D. Gerber, A. Natale, D. O'Dell, Virtual reality-based therapy for the treatment of balance deficits in patients receiving inpatient rehabilitation for traumatic brain injury, Disabil Rehabil. 2017 Jul;39(14):1380-1390. doi: 10.1080/09638288.2016.1195453.

research sites around the country. It is coordinated by the National Institute on Drug Abuse and the National Institute on Alcohol Abuse and Alcoholism and is supported by eight other NIH institutes and offices, as well as the Centers for Disease Control and Prevention and other federal partners.[113]

The ABCD study has already shown evidence of differences in the MRIs of the brains of 4,500 9- to 10-year-olds who were the heaviest users of electronic devices. Also, some children who spend two or more hours a day on screens had lower memory and language tests scores. What is still not known, however, is whether the time spent on digital devices is the primary cause of these effects. If not, what are the constellation of contributors? It may take years to answer some of these questions. An added challenge with electronic devices is that by the time some answers are proposed, new technology will come along to replace what we think we have come to better understand.

In Summary:

Electronic media is not intrinsically bad. People of all ages make good use of such technology. Electronic

---

[113] ABCD study completes enrollment, announces opportunities for scientific engagement. National Institutes of Health. (December 3, 2018.) [Accessed May 31, 2019.] https://www.nih.gov/news-events/news-releases/abcd-study-completes-enrollment-announces-opportunities-scientific-engagement

devices are extraordinary tools for communication, business and social exchange. The issue is how much gaming, emailing and Googling is too much for the mind, for a developing brain, as well as for a mature one. We need to discover our optimum level of brain stimulation at every stage of life. Although scientists have not yet determined the specific answers to these questions, we do know that a limited amount of electronic media may enrich minds and improve some forms of cognitive performance while too much can lead people to become unresponsive to the real world around them.

Pope Francis, in his 2016 World Communication Day Message said, "[T]he Internet...is something truly good, a gift from God," He also cautioned that the high-speed world of digital social media needed calm, reflection and tenderness if it was to be a "network not of wires but of people."

Regardless of how much people engage in electronic devices, to truly succeed in our lives, we need to learn to look into the eyes of another person and see someone worthy of love. We also need to raise our own eyes to receive the gaze of another upon us.

Section 7

The Need for Joy for Emotional Maturity

Saint Thomas states that there are 3 interior effects of charity[114]: joy, peace and mercy.

I would like to focus on joy as an essential ingredient to healthy growth and development. Consider for a moment that joy is like the temperature of an oven. We can choose our ingredients carefully, but the oven temperature will determine what our careful preparations will yield. Consider the effect of angry or joyful parents on a family and their prayer life. As joy increases, so does the possibility that human growth and development will go in a positive direction.

To build joy in our lives is not typically considered a spiritual practice and it is rarely considered a key factor for character development. Jesus, however, gave joy as a reason for his teaching in John 15:11 and the central features of his prayer for disciples in Jn 17:13.

Neuroscientists have demonstrated that the ventromedial prefrontal cortex, located above our eyes,

---

[114] Saint Thomas Aquinas, Summa Theologiae, II, II, ques. 28-30.

has a very important relational skill of figuring out the least harmful solution to every situation. It is responsible for judging, planning and reflection. This damage-control system in our brain needs to be trained with a full range of relational skills that affect us interiorly with joy.

From a brain perspective, joy stimulates the growth of the brain systems involved in character formation, identity consolidation and moral behavior. Joy is a "glad to be together." Joy is relational. Joy is found in smiles, play and love. Joy can be equally powerful when we are in painful states. We feel very keenly if there is anyone who is glad to be with us when we are hurting. When we settle into the arms of a friend who rushed to the emergency room while we waited to see whether a loved one would survive, we weep with relief rather than bounce with euphoria, but it is joy all the same. Someone is with us, and we are not alone. Joy then lets us "rejoice with those who rejoice and weep with those who weep," for we are deeply united. Joy, like any powerful internal drive, can be combined with other experiences to provide many flavors, but the signature of joy is that we are sharing the moment with someone who is glad we are there.

Since joy is relational, the longer a disruption in our relationships lasts the less our joy becomes. Spending more than a few seconds in very cold water can be deadly, but it does not matter how cold the water is if you do not stay in very long. The secret to staying warm is to reduce the heat loss and exposure to the cold. The skill of swimming helps us get out of the cold water quickly—although at first this skill seems unrelated to heat loss. To retain joy, we need the skills to return to joy quickly.

We must practice keeping our minds in a relational mode where the relationship is always more important than the problem. We must be trained to keep love in first place while we are in pain, upset and facing problems—particularly problems caused by the people in our lives. A baby will simply scream with no thought for the relationship with others every time there is a problem. While no one blames the baby for this type of reaction, no one wants that child to grow up with no other skills for problem-solving or relationship awareness than screaming.

American Christianity has grown up under the influence of voluntarist philosophy. Simply put, voluntarism believes that the will is the highest and strongest function of the person, higher even than emotions or the intellect.

From a brain perspective, the will is a neurologically weak factor. The will is a fickle cortical function that starts to disappear as soon as we are a little sleepy. The will is well-down the brain's control hierarchy for making changes in character or identity. It is wired to have only a weak influence. Intentions are some of the first things to fade under strain from a brain that receives low joy messages. Under the chemical effect of drinking two beers, many people forget their intention to stop drinking after three beers. Intense emotions, fatigue, novelty and many other factors also derail the best intentions of the low-joy brain. When people do not do what they intended, they display what we generally consider a lack of good character. The voluntarist solution is to renew the vision and intentions to do better next time.

Voluntarists took their lead from medieval notions of human nature and assumed that humans have only one will. Both Scripture and the brain suggest different parts of us have different wills. For example, Saint Paul in Romans (7: 19) states, "For I do not do the good I want to do, but the evil I do not want to do." What one part of the brain wills is not necessarily also willed by other parts of the brain that may have contrary urges, motives and choices. The brain is designed to unify its various wills under one identity, provided joyful relationships are the norm. Mature members of the community help to model how to stay relational during upsetting emotions.

It is also true that the brain does not like to act and reflect simultaneously; this is one of the reasons we need solitude at times. A mature identity is needed to adjudicate the two. This mature identity only forms in the presence of joy. Even when our identity is strong, character change is extremely difficult to achieve from the "will" end of the brain. The will is more the result of the brain's processes than the cause of those processes. The will is where processing ends rather than begins. By contrast, brain processing that leads to identity and character change begins in the love and attachment regions. Even the motivation to change is birthed by love and attachment. Joy-based character change always moves us in the direction of being more like the One we love.

Scientific studies have shown that persons who were highly motivated and who were offered the same spiritual and psychological help did not all achieve or sustain positive changes of character. Those who lacked relationships with people who were glad to be with them did poorly while others prospered to the extent they had joyful relationships.

Joy activates the brain's social engagement system and prepares us to engage with God and others. Because the brain's development is relational, relational activities are best suited for human and spiritual formation that transforms the mind.

We already know that wounded relational experiences create barriers to formation. To be effective as a means for conversion, joy and relational skills need to be intentionally introduced for the healing of those who lack relational joy.

Face-to-face interactions are becoming rare:

Every child entering the world starts with a near-total absence of relational ability. Where even sixty years ago, children spent virtually all their time interacting with others, playing and comparing abilities, they now spend the majority of their time focused on machines that have no relational skills or awareness of the child's presence. Babies are watching television; they have movies in preschool, day care, church, at home and with baby-sitters.

Television, computers, movies, books or video games cannot spread relational skills. Brain requirements for successful transmission of joy skills do not propagate through media, Internet, webcams or even telecommunications. We cannot raise babies to be joyful human beings via the Internet or television.

What can we expect as relational skills decline?

As relational joy skills decrease globally, we can expect a long series of shifts in how people relate. These shifts will remain invisible to the generation where the shift happens. For them, "this is just how people are." Lower joy skills will always mean increased violence and predatory behavior.

Here is what we may expect with lower joy-skill levels:

- functionality replaces relationship
- managing problems replaces restoring joy
- schools manage problems rather than raise human children
- self-justified behavior is the norm
- pseudo-identity becomes the ideal
- narcissism increases
- unresolved conflicts increase
- addictions increase

The great hope of claiming and reclaiming relational joy

Recall that the prefrontal cortex has the very important relational skill of figuring out the least harmful solution to every situation. This damage control system in our brain needs to be trained with the full range of joy skills. Part of damage control is our ability to return to joy. Returning to joy is how we go about saving our relationships when others are not glad to be with us. Our brain's identity center in the prefrontal cortex considers the examples we have seen in our families and life experiences and tries to figure out the least painful solution. If God is always here with us, an interactive awareness of God's presence will help us acquire a new response. We can acquire these skills from God because Jesus will redemptively bypass the gaps left by our families and life experiences and teach us the way back to joy so that we can then practice these skills with others. This is the gift of community.

Joy and conversion

The brain is more deeply changed by whom it loves (who brings me joy) than by what it thinks. Perhaps this is the real reason we have not seen the connection between human and spiritual formation and brain science. Conversion is more about more intimately receiving God's love for us than right thinking. Too often we forget that the deepest brain change comes from our loves and not our thoughts. I am not suggesting that we abandon right thinking, much less right vision or intention, but rather that we consider adding joy to the vision, intention and means.

For example, spiritual formation practices will be very different depending on whether the mind grows into the image of Christ primarily because of what one believes or because of whom one loves. The first approach exercises the will and corrects thoughts. The second approach focuses on removing barriers to love. If conversion is about forming a new attachment with God built on love and joy, then who becomes our joy will determine who forms our character—perhaps even whether we will change character at all.

Where joy is low, the means of conversion must produce a rapid increase in joy as we become more eager to be together with God and others. In fact, our eagerness must extend beyond those who are being converted with us to include those in need of conversion. Joy spreads. In *Joy of the Gospel*, Pope Francis articulates time and time again that joy is the way the gospel should be spread.

Appendix A

Discussion Booklet for Parents

https://youtu.be/Thkl8-q2pYc

Welcome to the CMA University Video Vignette Group Discussion—"Why Can't You Put Down that Phone?"

Over the course of your discussions you will view a brief video (approximately 5½ minutes), which will highlight some important principles about screen media use.

After the video presentation, you will engage in a group discussion. You will have the opportunity to talk about how screen time influences your daily life. You'll hear about some ideas that have benefitted many other young adults.

Some groups might want to schedule additional discussion sessions to share their experiences and ideas. Enjoy!

Section 1

Video Review Questions

- Did you hear anything that surprised you in the video?
- What does Marshall McLuhan mean when he says: "It's the medium more than the content that is the message"?
- Why is unpredictability so powerful in stimulating desire and craving?
- How does operant conditioning or cueing make you want to check your cell phone?
- Persons who excessively engage in online gaming, for example, *World of Warcraft* or *Minecraft*, can have trouble stopping. Describe why this happens.
- What is your response to hearing that: Steve Jobs, founder of the iPhone, never permitted his kids to use an iPhone; and that Evan Williams, founder of Blogger, Twitter and Medium, refused to give his two young sons an iPad and instead bought them hundreds of books?[115]

---

[115] N. Bilton, Steve Jobs was a low-tech parent. New York Times. (September 10, 2014.) (Accessed on May 31, 2019.) https://www.nytimes.com/2014/09/11/fashion/steve-jobs-apple-was-a-low-tech-parent.html

Group Discussion Questions

The following questions are intended to help you reflect on your child's use of screen media.

- In what ways do social media platforms/ texting benefit you, your children, family and friends?
- How much does your child's social interactions take place on screen media?
- How much screen time does your child have per weekday and per weekend?
- Does your child talk mostly about online gaming, social media or texting friends?
- What family activities have been interfered with by your child's use of electronic devices?
- Does your child multi-task when using his or her screens? What does your child choose to do when multi-tasking?
- Is using an electronic device the first thing he or she asks to do when home from school/on weekends?
- How much daily physical exercise is your child getting?
- Does your child take his or her cell phone/tablet to bed? How has this affected the amount of sleep your child gets at night?
- How does electronic media affect your child's mood?
- In what situations are you tempted to give your child an electronic device as a distraction?

- What guidelines have you established for your child's use of electronic media?
- Have you had discussions with your child about appropriate use of cell phones, tablets, and computers? If so, what was fruitful about the conversations and what was challenging about the conversations?
- What concerns do you have about limiting your child's use of electronic devices?
- Has your child ever experienced FOMO (Fear Of Missing Out)? In what situations?
- Are you concerned about taking away a screen activity that will make your child "not fit in" with his or her peers?
- What skills or talents could your child develop if he or she were spending an hour less using electronic devices each day?

Section 2

Tips and Suggestions

It is difficult to know how much screen media is healthy and safe. Medical professionals offer some of the following guidelines that can be adapted for the needs of each of your children:

- Children younger than 1 1/2 to 2 years: Avoid media use (except for video chatting with family and loved ones).

- Preschool children: No more than one hour per day with parental screening.
- Grade-schoolers and teens: Don't let media displace other important activities. Aim for one hour of daily exercise, media-free meals, "unplugged" down-time and a full night's sleep.
- Be your children's media mentor: View media with your child (enjoy playing, sharing and teaching), to know what your child is viewing and guide appropriately. Also, model healthy screen-use habits yourself.

The good news is that the problematic effects of screen overuse can be addressed. There is much we can do to restore the balance that existed before the age of cell phones, tablets, emails, social networking and on-demand viewing.

We know the tricks that advertisers use to tempt us—a mouthwatering sundae or a shiny red car. Electronic design engineers also know how to get our attention. The color red in notifications on our cell phones or tablets triggers a desire to check what it is all about. Auto-play in videos can get us hooked to watching much longer than we intended. More and more cell phone and tablet functions are designed to get us to spend more time using these devices.

What would a healthy screen time plan look like for you, your family and in your interactions with your friends?

Here are a few suggestions:

Say no to notifications. A notification is intended to let you know something important needs attention. Most phone and tablet notifications are machine-automated and do not involve people. They are intended to get you to engage with an app you might not have otherwise thought about. Set notifications for things that are important to you. For example, you might want notifications when you receive a message from your child, spouse or supervisor.

To do so, on an iPhone go to Settings/Notifications/Turn off everything except your message apps or other important tools.

Put your phone on "do not disturb" mode and allow only calls and messages from "favorites" to whom you have a responsibility. You might be surprised how much time interruptions from messages take away from what is important to you.

Remove social media and other apps from your phone. Tapping on an app, such as Facebook or Twitter, is easy to do—without much thinking. Limit your access to these apps to your home computer. If you have the urge to check your social media more often, ask yourself why: are you bored, lonely, anxious, bothered or concerned? See if this need could be addressed more effectively in a real-life interaction or activity.

Keep only the apps on your home screen that are necessary. For example, some important apps to keep on your home screen may be your contacts, text messaging, email, maps and calendar apps that you use daily for family, work and select friends. Move your other apps to your second and third screens; what you do not see right away when you pick up your phone or tablet will be less tempting to access.

No screen zones. Use good screen etiquette. No screens at mealtime will help facilitate face-to-face conversations and family bonding. Developing speech and language skills is strongly linked to developing human bonding, empathy, social responsibility, as well as strengthening thinking, reading and writing ability. These essential human skills, and most especially you, cannot be replaced by technology. Also keep cell phones, tablets and laptops out of the bedroom at night. This will promote more minutes of sleep and better-quality sleep. Of course, no one should read or send messages while driving.

Use an old-fashioned alarm clock. Not using the phone or tablet as an alarm clock will prevent you and your children from being tempted to engage in the many other functions the phone or tablet offers at a time and place designated for rest and sleep.

Help your child modify his or her screen habits. The goal of limiting screen time is not simply to be rid of the problematic effects but to optimize human development and maturation. Healthy well-balanced child development takes place primarily through your face-to-face interactions. Again, your important role as a parent cannot be replaced by technology.

Since some screen time is all but inevitable in our world today, it is far easier to help children develop good habits in the first place than to correct existing bad habits.

Developing Positive Screen Habits

Encourage children to connect what they see in the screen world to their experience of the real world.

For example, if an app asks a child to sort wooden blocks by color, ask the child to sort real wooden blocks by color or to sort his or her clothes by color.

Encourage real world interpersonal interactive activities: board games, sports, hiking, outdoor/ indoor work projects and turn-taking conversations.

Invite children to explain what they think is happening in a game or in an app they are using. Have them point to and identify the characters and their thoughts about the characters' behaviors. Have them comment on what behaviors might be modified and why.

The process of human growth and development also involves learning appropriate decision-making skills and social etiquette. Teaching children the virtues of moderation, kindness, generosity and empathy, for example, will go a long way in helping them become successful adults.

What Drives Your Child's Use of Screens?

Help your child unpack what your child sees as the benefits of screen usage to understand the underlying need. For example, if a teenager checks her Instagram or Snapchat account many times a day, she likely wants to feel connected to her friends.

Help in addressing this need to mutually explore other means of feeling connected and in which she feels validated in the absence of those "likes."

Remember, what works for one person may not work for another. Other examples to possibly consider:

A child who is bored and turns to screen usage might benefit from joining a sports team or drama club.

The vulnerability a bullied gamer experiences might be met, in part, by martial arts classes, in addition to speaking with his parents.

Once the underlying motivation for screen use is better understood, a new routine that satisfies this underlying motive can be sought.

Your efforts to better understand your child's motivations will more effectively address the underlying psychological need.

Modifying Your Child's Screen Habits

It is important to identify your concerns about your child's screen use. Then set goals to address these concerns. Setting clear goals helps keep everyone in your family focused on what is important. Without clear, relevant goals results are often disappointing.

Consult your experience. Ask your child's teacher, sitter, or grandparent as to what difficulties they observe in your child's screen usage. Is your child uncooperative, not interested in schoolwork or reluctant to participate in family activities? Focus on two or three behaviors that are most concerning and state your goal(s).

Consider using the acronym SMART goal which refers to a Specific, Measurable, Attainable, Relevant and Trackable goal for which you develop an action plan to address.

Let's say your child does not want to go to bed because he or she wants to finish an online game/ social media post and often is sleepy and does not feel rested during the day. A SMART goal would be to work on getting better quality sleep for the next week.

- Specific-Get more hours of sleep for your child
- Measurable-Note the time bedroom lights are turned off and the time your child got up
- Attainable-Yes, your child can get more sleep when not using cell phone, tablet, and laptop in bed
- Relevant-Not bringing the cell phone, tablet, and laptop into the bedroom improves quality of sleep
- Trackable-Track how many days sleep quality is improved

Next assess the cost and benefit of this SMART goal action plan for your child. Decide whether you want to continue with this action plan, modify it, or completely change it and why. Write this information down in your journal.

For example, ask your child his or her experience of sleep without taking his or her cell phone/tablet/ laptop to bed. Share your experience with your child in his or her having done so. Depending on the age of your child, invite him or her to draw a picture or write down the results of going to bed without any device around.

Develop SMART goal action plans for your other areas of concern.

To provide another example of a SMART goal action plan, a parent notices that the child complains or cries when the use of an electronic device is limited. A SMART goal may be to reduce the frequency of this behavior by replacing device time with one-on-one time to build a toy model plane, to make jewelry, or to play with Legos, for example.

Help Replace Screen Time with Positive Family Activities

Consider scheduling a range of activities for the whole family, as well as one-on-one time with each child that communicate "I want to be with you; you are important to me; and we can have fun together." Be creative and engage your personal interests for non-screen time fun. For example:

- Take a family walk
- Have a picnic
- Ride bikes together
- Play Frisbee
- Play catch, jump rope
- Cook a meal together
- Plant a garden
- Go out for an ice cream treat
- For very young children

- Teach them nursery rhymes
- Play Simon says, Peek-a-boo or London bridges falling down
- You can also ask your grand/parents what they did before the Internet for some screen-free fun.

Parents need to be good models for their children. The bonus is that your child's executive functioning, that is the work of his or her part of the brain responsible for reflection, integration of thoughts and planning will be enhanced. Your child's pleasure center will also be less stimulated, and this will contribute to your child's greater sense of calm.

Utilize Apps to Help Increase Screen-free Time

Many apps are available to help you be more mindful of your family members' use of screens and to allow for more screen-free time. One example is Onward, an app that helps block websites and apps and tracks phone or tablet usage.

There are also apps that allow you to designate screen-free time for the whole family. Some examples include:

- Moment Family
- Breakfree
- DinnerTime Plus

Consider Internet firewalls and filtering systems for your cell phone, tablet, and computer as appropriate for your family. There are many on the market. Check out what system is most appropriate for you and your family. Some suggestions include:

- Circle Go
- Qustodio
- Covenant Eyes
- Mobicip
- OurPact
- App blocking devices to consider:
- Freedom for Apple and Windows products
- Offtime for Android products
- GizmoPal is a GPS tracking watch for kids that may be useful for your family.

Challenge yourself to reduce screen time

Initially, try to reduce your screen time by one hour per week.

You could also challenge your family to a no-screen day. Having your family give up screens for a day is not intended to be a punishment. It is an opportunity for your family to reflect on the ways in which screen activity separated you and ways in which you decide you want to enrich your time together.

Get your family on board.

Choose a day that is realistic to be screen-free.

Take some time to reflect on your screen-free day as a family.

What did you notice?

How did your no-screen day compare with a day when using screens?

What do you notice when you were interacting face-to-face with others for a day?

What was your family able to do because you were not engaged with your individual screens?

Based upon your experience with a no-screen day, consider finding an entire weekend to go screen-free.

Need help?

For some persons, online gaming or other Internet activities become an irresistible obsession. Help is available. Treatment options range from limited out-patient therapy to intensive residential inpatient programs. Check online for treatment options that fit you and your family's needs.

Group Facilitator's Guide

Tips to help you plan

- A group facilitator does not have to be an expert in neurophysiology or psychology. Rather, a facilitator makes others feel welcome and is able to draw out those who need a little extra encouragement; and reins in the more exuberant participants.
- Familiarize yourself with the video, questions, exercises and suggested resources before your meeting. Spend time considering how people

may respond to them. This will help you to be prepared for some directions the discussion might take.

- Schedule a 60-minute discussion session for your group as far in advance as possible and publicize it to your target audience. Because there is more material in the booklet than can be covered in an hour, you might select questions you want your group to focus on during the hour. Let participants know that also listed in the booklet are helpful tips and suggestions are that they can consider on their own.
- Some groups might want to meet two or three times.

  - During the first session, participants can focus on questions of interest.
  - For the second session, invite participants to share some tips or recommendations that they implemented.
  - If a third discussion session is desired, focus on particular areas of interest and encourage ongoing implementation of tips, recommendations or other helpful ideas that emerged during the group discussion.

- Arrange for an appropriate meeting space and for A/V equipment to play the video "Why can't you put down that phone?"

- Once you know how many people will be attending the discussion, make copies of the discussion booklet.
- Consider a hospitality table with simple refreshments, which may help to break the ice and foster discussion within the group.
- Arrive early to the session to set up the registration table and hospitality area; arrange the gathering space for the video presentation and discussion.
- Consider praying for God's blessing upon all those who will participate in your discussions.
- Consider choosing a prayer to begin each discussion session.
- The most important preparation for you as a facilitator is to go through the questions yourself. Your personal participation in the study will help you relate to and understand the experiences of your group members.
- Remember, it is not the facilitator's job to answer every question that comes up. Whenever possible, ask the participant what he or she thinks first. If appropriate, engage the others present for their thoughts.
- Stay on time and on topic. Some persons find it helpful to have a second person in the group who is responsible for giving a ten minute "warning" signal to allow ample time to finish the discussion and perhaps close with a prayer. Finishing the discussion period promptly is essential and shows respect for the participants.

- The group will expect you to keep the discussion moving.
    - Watch for clues that a timid person has something to say and encourage that person without putting him or her on the spot.
    - Kindly “rein in” participants who dominate the group. If they continue, ask them privately to help you get others to participate.
    - Gently redirect tangential remarks or questions.
- Don’t share confidences outside the group.
- Enjoy yourself!

How to structure each discussion session

Here is a simple agenda:

- Introduction (5–10 minutes) for gathering and opening remarks
- View video “Why Can’t You Put Down that Phone?” (5 ½ minutes)
- Group discussion (30 minutes) engaging participants using the booklet with its discussion questions, tips and suggestions and resources
- Closing remarks and possible prayer (5–10 minutes)

Section 1

Video Review Questions

1. Did you hear anything that surprised you in the video?
2. What does Marshall McLuhan mean when he says: "It's the medium more than the content that is the message"?
3. Why is *unpredictability* so powerful in stimulating desire and craving?
4. How does *operant conditioning* or *cueing* make you want to check your cell phone?
5. Persons who excessively engage in online gaming, for example, *World of Warcraft* or *Minecraft*, can have trouble stopping. Describe why this happens.
6. What is your response to hearing that: Steve Jobs, founder of the iPhone, never permitted his kids to use an iPhone and that Evan Williams, founder of Blogger, Twitter and Medium, refused to give his two young sons an iPad and instead bought them hundreds of books?[116]

[116] N. Bilton, Steve Jobs was a low-tech parent. New York Times. (September 10, 2014.) (Accessed on May 31, 2019.) https://www.nytimes.com/2014/09/11/fashion/steve-jobs-apple-was-a-low-tech-parent.html

Group Discussion Questions

The following questions are intended to help you engage groups to reflect on their child's or children's use of screen media.

1. In what ways do social media platforms/ texting benefit you, your children, family and friends?
2. How much does your child's social interactions take place on screen media?
3. How much screen time does your child have per weekday and per weekend?
4. Does your child talk mostly about online gaming, social media or texting friends?
5. What family activities have been interfered with by your child's use of electronic devices?
6. Does your child multi-task when using his or her screens? What does your child choose to do when multi-tasking?
7. Is using an electronic device the first thing he or she asks to do when home from school/on weekends?
8. How much daily physical exercise is your child getting?
9. Does your child take his or her cell phone/ tablet to bed? How has this affected the amount of sleep your child gets at night?
10. How does electronic media affect your child's mood?
11. In what situations are you tempted to give your child an electronic device as a distraction?

12. What guidelines have you established for your child's use of electronic media?
13. Have you had discussions with your child about appropriate use of cell phones, tablets, and computers? If so, what was fruitful about the conversations and what was challenging about the conversations?
14. What concerns do you have about limiting your child's use of electronic devices?
15. Has your child ever experienced FOMO (Fear Of Missing Out)? In what situations?
16. Are you concerned about taking away a screen activity that will make your child "not fit in" with his or her peers?
17. What skills or talents could your child develop if he or she were spending an hour less using electronic devices each day?

Section 2

Tips and Suggestions

It is difficult to know how much screen media is healthy and safe. Medical professionals offer some of the following guidelines that can be adapted for the needs of each of your children:

- Children younger than 1 1/2 to 2 years: Avoid media use (except for video chatting with family and loved ones).
- Preschool children: No more than one hour per day with parental screening.

- Grade-schoolers and teens: Don't let media displace other important activities. Aim for one hour of daily exercise, media-free meals, "unplugged" down-time and a full night's sleep.
- Be your children's media mentor: View media with your child (enjoy playing, sharing and teaching), to know what your child is viewing and guide appropriately. Also, model healthy screen-use habits yourself.

The good news is that the problematic effects of screen overuse can be addressed. There is much we can do to restore the balance that existed before the age of cell phones, tablets, emails, social networking and on-demand viewing.

We know the tricks that advertisers use to tempt us—a mouthwatering sundae or a shiny red car. Electronic design engineers also know how to get our attention. The color red in notifications on our cell phones or tablets triggers a desire to check what it is all about. Auto-play in videos can get us hooked to watching much longer than we intended. More and more cell phone and tablet functions are designed to get us to spend more time using these devices.

What would a healthy screen time plan look like for you, your family and in your interactions with your friends?

Here are a few suggestions:

Say no to notifications. A notification is intended to let you know something important needs attention. Most phone and tablet notifications are machine-automated and do not involve people. They are intended to get you to engage with an app you might not have otherwise thought about. Set notifications for things that are important to you. For example, you might want notifications when you receive a message from your child, spouse or supervisor.

- To do so, on an iPhone go to Settings/Notifications/Turn off everything except your message apps or other important tools.

Put your phone on "do not disturb" mode and allow only calls and messages from "favorites" to whom you have a responsibility. You might be surprised how much time interruptions from messages take away from what is important to you.

Remove social media and other apps from your phone. Tapping on an app, such as *Facebook* or *Twitter*, is easy to do—without much thinking. Limit your access to these apps to your home computer. If you have the urge to check your social media more often, ask yourself why: are you bored, lonely, anxious, bothered, or concerned? See if this need could be addressed more effectively in a real-life interaction or activity.

Keep only the apps on your home screen that are necessary. For example, some important apps to keep on your home screen may be your contacts, text messaging, email, maps and calendar apps that you use daily for family, work and select friends. Move your other apps to your second and third screens; what you do not see right away when you pick up your phone or tablet will be less tempting to access.

No screen zones. Use good screen etiquette. No screens at mealtime will help facilitate face-to-face conversations and family bonding. Developing speech and language skills is strongly linked to developing human bonding, empathy, social responsibility, as well as to thinking, reading and writing ability. These essential human skills, and most especially you, cannot be replaced by technology. Also keep cell phones, tablets, and laptops out of the bedroom at night. This will promote more minutes of sleep and better-quality sleep. Of course, no one should read or send messages while driving.

Use an old-fashioned alarm clock. Not using the phone or tablet as an alarm clock will prevent you and your children from being tempted to engage in the many other functions the phone or tablet offers at a time and place designated for rest and sleep.

Help your child modify his or her screen habits. The goal of limiting screen time is not simply to be rid of the problematic effects but to optimize human development and maturation. Healthy well-balanced child development takes place primarily through your face-to-face interactions. Again, your important role as a parent cannot be replaced by technology.

Since some screen time is all but inevitable in our world today, it is far easier to help children develop good habits in the first place than to correct existing bad habits.

Developing Positive Screen Habits

Encourage children to connect what they see in the screen world to their experience of the real world.

- For example, if an app asks a child to sort wooden blocks by color, ask the child to sort real wooden blocks by color or to sort his or her clothes by color.
- Encourage real world interpersonal interactive activities: board games, sports, hiking, outdoor/indoor work projects, and turn-taking conversations.
- Invite children to explain what they think is happening in a game or in an app they are using. Have them point to and identify the characters and their thoughts about the characters' behaviors. Have them comment on what behaviors might be modified and why.

The process of human growth and development also involves learning appropriate decision-making skills and social etiquette. Teaching children the virtues of moderation, kindness, generosity and empathy, for example, will go a long way in helping them become successful adults.

What Drives Your Child's Use of Screens?

Help your child unpack what your child sees as the benefits of screen usage to understand the underlying need.

- For example, if a teenager checks her *Instagram* or *Snapchat* account many times a day, she likely wants to feel connected to her friends.
- Offer assistance in addressing this need to mutually explore other means of feeling connected and in which she feels validated in the absence of those "likes."

Remember, what works for one person may not work for another. Other examples to possibly consider:

- A child who is bored and turns to screen usage might benefit from joining a sports team or drama club.
- The vulnerability a bullied gamer experiences might be met, in part, by martial arts classes, in addition to speaking with his parents.

Once the underlying motivation for screen use is better understood, a new routine that satisfies this underlying motive can be sought. Your efforts to better understand your child's motivations will more effectively address the underlying psychological need.

Modifying Your Child's Screen Habits

It is important to identify your concerns about your child's screen use. Then set goals to address these concerns. Setting clear goals helps keep everyone in your family focused on what is important. Without clear, relevant goals results are often disappointing. Consult your experience. Ask your child's teacher, sitter, or grandparent as to what difficulties they observe in your child's screen usage. Is your child uncooperative, not interested in schoolwork or reluctant to participate in family activities? Focus on two or three behaviors that are most concerning and state your goal(s).

Consider using the acronym SMART goal which refers to a Specific, Measurable, Attainable, Relevant and Trackable goal for which you develop an action plan to address.

Let's say your child does not want to go to bed because he or she wants to finish an online game/ social media post and often is sleepy and does not feel rested during the day. A SMART goal would be to work on getting better quality sleep for the next week.

- Specific-Get more hours of sleep for your child
- Measurable-Note the time bedroom lights are turned off and the time your child got up
- Attainable-Yes, your child can get more sleep when not using cell phone, tablet, and laptop in bed

- Relevant-Not bringing the cell phone, tablet, and laptop into the bedroom improves quality of sleep
- Trackable-Track how many days sleep quality is improved

Next assess the cost and benefit of this SMART goal action plan for your child. Decide whether you want to continue with this action plan, modify it, or completely change it and why. Write this information down in your journal.

For example, ask your child his or her experience of sleep without taking his or her cell phone/tablet/ laptop to bed. Share your experience with your child in his or her having done so. Depending on the age of your child, invite him or her to draw a picture or write down the results of going to bed without any device around.

Develop SMART goal action plans for your other areas of concern. To provide another example of a SMART goal action plan, a parent notices that the child complains or cries when the use of an electronic device is limited. A SMART goal may be to reduce the frequency of this behavior by replacing device time with one-on-one time to build a toy model plane, to make jewelry, or to play with Legos, for example.

Help Replace Screen Time with Positive Family Activities

Consider scheduling a range of activities for the whole family, as well as one-on-one time with each child that communicate "I want to be with you; you are important to me; and we can have fun together." Be creative and engage your personal interests for non-screen time fun. For example:

- Take a family walk
- Have a picnic
- Ride bikes together
- Play Frisbee
- Play catch, jump rope
- Cook a meal together
- Plant a garden
- Go out for an ice cream treat
- For very young children

    - Teach them nursery rhymes
    - Play *Simon says, Peek-a-boo* or *London bridges falling down*

You can also ask your grand/parents what they did before the Internet for some screen-free fun.

Parents need to be good models for their children. The bonus is that your child's executive functioning, that is the work of his or her part of the brain responsible for reflection, integration of thoughts and planning will be enhanced. Your child's *pleasure center* will also be less stimulated, and this will contribute to your child's greater sense of calm.

Utilize Apps to Help Increase Screen-free Time

Many apps are available to help you be more mindful of your family members' use of screens and to allow for more screen-free time. One example is *Onward*, an app that helps block websites and apps and tracks phone or tablet usage.

There are also apps that allow you to designate screen-free time for the whole family. Some examples include:

- *Moment Family*
- *Breakfree*
- *DinnerTime Plus*

Consider cell phone, tablet, and computer Internet firewalls and filtering systems as appropriate for your family. There are many on the market. Check out what system is most appropriate for you and your family. Some suggestions include:

- *Circle Go*
- *Qustodio*
- *Covenant Eyes*
- *Mobicip*
- *OurPact*

App blocking devices to consider:

- *Freedom* for Apple and Windows products
- *Offtime* for Android products

GizmoPal is a GPS tracking watch for kids that may be useful for your family.

Challenge yourself to reduce screen time.

Initially, try to reduce your screen time by one hour per week.

You could also challenge your family to a no-screen day. Having your family give up screens for a day is not intended to be a punishment. It is an opportunity for your family to reflect on the ways in which screen activity separated you and ways in which you decide you want to enrich your time together.

- Get your family on board.
- Choose a day that is realistic to be screen-free.
- Take some time to reflect on your screen-free day as a family.

    - What did you notice?
    - How did your no-screen day compare with a day when using screens?
    - What do you notice when you were interacting face-to-face with others for a day?
    - What was your family able to do because you were not engaged with your individual screens?

Based upon your experience with a no-screen day, consider finding an entire weekend to go screen-free.

Need help?

For some persons, online gaming or other Internet activities become an irresistible obsession. Help is available. Treatment options range from limited outpatient therapy to intensive residential inpatient programs. Check online for treatment options that fit you and your family's needs.

Appendix B

Discussion Booklet for Young Adults

https://youtu.be/Thkl8-q2pYc

Welcome to the CMA University Video Vignette Group Discussion—"Why Can't You Put Down that Phone?" Over the course of your discussions you will view a brief video (approximately 5 ½ minutes) which will highlight some important principles about screen media use. After the video presentation, you will engage in a group discussion. You will have the opportunity to talk about how screen time influences your daily life. You will hear some ideas that have benefitted many other young adults. Some groups might want to schedule additional discussion sessions to share their experiences and ideas. Enjoy!

Section 1

Video Review Questions

1. Did you hear anything that surprised you in the video?
2. What does Marshall McLuhan mean when he says: “It’s the medium more than the content that is the message”?
3. Why is *unpredictability* so powerful in stimulating desire and craving?
4. How does *operant conditioning* or *cueing* make you want to check your cell phone?
5. Persons who excessively engage in online gaming, for example, *World of Warcraft* or *Minecraft*, can have trouble stopping. Describe why this happens.
6. What is your response to hearing that: Steve Jobs, founder of the iPhone, never permitted his kids to use an iPhone and that Evan Williams, founder of Blogger, Twitter and Medium, refused to give his two young sons an iPad and instead bought them hundreds of books?[117]

[117] N. Bilton, Steve Jobs was a low-tech parent. New York Times. (September 10, 2014.) (Accessed on May 31, 2019.) https://www.nytimes.com/2014/09/11/fashion/steve-jobs-apple-was-a-low-tech-parent.html

Group Discussion Questions

The following questions are intended to help you reflect on your use of screen media.

1. In what ways do social media platforms and texting benefit you, your family, and friends?
2. What is your reaction when you pick up your phone to see if you have any messages or "likes" and there aren't any? How does this make you feel?
3. Do you catch yourself accessing social media when you should be doing something else?
4. How much time do you spend with friends online as compared with friends in real life?
5. In what situations do you interrupt a conversation you are having with someone face-to-face to check a message you received?
6. Do you multi-task when using your screens? What do you choose to do when multi-tasking?
7. Have you ever experienced FOMO? In what situations?
8. How much personal time do you daily spend on your electronic device(s)?
9. Are there triggers for when you use your electronic device? When bored? When alone? When you first awake? When waiting in line? Others?
10. How does your use of electronic devices affect your mood?
11. When you use your electronic device, how often do you have a clear purpose in mind?

12. Do you take your phone to bed? How long are you on your phone in bed before trying to all asleep?
13. What times or situations would you think it is inappropriate to engage in social media?
14. What would concern you if you were asked to turn off your devices for a day?
15. What skills or talents could you develop if you were spending an hour less on social media each day?

Section 2

Be a Smarter Screen User. We know the tricks that advertisers use to tempt us—a mouthwatering sundae or a shiny red car. Electronic design engineers also know how to get our attention. The color red in notifications on our cell phones and tablets triggers a desire to see what it is all about. Auto-play in videos can get us hooked to watching much longer than we intended. More and more cell phone and tablet functions are designed to get us to spend more time using these devices.

Here are a few suggestions to help you be more aware of your tech usage:

Say no to notifications. A notification is intended to let you know something important needs attention. Most phone and tablet notifications are machine-automated and do not involve people. They are intended to get you to engage with an app you might not have otherwise thought about. Set notifications only for things that are important to you.

Put your phone on "do not disturb" mode and allow only messages from "favorites." You might be surprised how much time interruptions from messages take away from what is important to you.

No screen zones. Use good screen etiquette. No screens at mealtime will help facilitate better face-to-face conversations. Also keep your phone and tablet out of your bedroom at night. You will probably get more minutes of sleep and better-quality sleep. Of course, do not read or send messages while driving.

Use an old-fashioned alarm clock. Not using your phone or tablet as an alarm clock will prevent you from being tempted to engage in the many other functions your phone and tablet offers you at a time and place designated for rest and sleep.

Utilize Apps to Help Increase Screen-free Time
Many apps are available to help you be more mindful of your use of screens and to allow for more screen-free time. One example is *Onward*, an app that helps block websites and apps and tracks phone or tablet usage. There are also apps that help you to designate screen-free times during the day. Some examples include:

- *Moment Family*
- *Breakfree*

Consider Internet firewalls and filtering systems for you cell phone, tablet, and computer. There are many on the market. Some suggestions include:

- *Circle Go*
- *Qustodio*
- *Covenant Eyes*

App blocking devices to consider:

- *Freedom* for Apple and Windows products
- *Offtime* for Android products

Challenge yourself to reduce screen time. Initially, try to reduce your screen time by one hour per week.

You could also challenge yourself to a no-screen day. Giving up screens for a day is not intended to be a punishment. It is an opportunity for you to challenge yourself to interact with people in new ways.

- Get your friends on board.
- Choose a day that is realistic to be screen-free.
- Take some time to reflect on your screen-free day.
  - What did you notice?
  - How did your no-screen day compare with a day when using screens?
  - What do you notice when you were interacting face-to-face with others for a day?

- What were you able to do because you were not engaged with your screens?

Based upon your experience with a no-screen day, consider finding an entire weekend to go screen-free.

Need help?

For some persons, online gaming or other Internet activities become an irresistible obsession. Help is available. Treatment options range from limited outpatient therapy to intensive inpatient programs. Check online for treatment options that fit your needs.

Group Facilitator's Guide

Tips to help you plan

- A group facilitator does not have to be an expert in neurophysiology or psychology. Rather, a facilitator makes others feel welcome and is able to draw out those who need a little extra encouragement; and reins in the more exuberant participants.
- Familiarize yourself with the video, questions, exercises and suggested resources before your meeting. Spend time considering how people may respond to them. This will help you to be prepared for some directions the discussion might take.
- Schedule a 60-minute discussion session for your particular group as far in advance as

possible and publicize it to your target audience. Because there is more material in the booklet than can be covered in an hour, you might select questions you want your group to focus on during the hour. Let participants know that also listed in the booklet are helpful tips and suggestions are that they can consider on their own.

- Some groups might want to meet two or three times.

    - During the first session, participants can focus on questions of interest.
    - For the second session, invite participants to share some tips or recommendations that they implemented.
    - If a third discussion session is desired, focus on particular areas of interest and encourage ongoing implementation of tips, recommendations or other helpful ideas that emerged during the group discussion.

- Arrange for an appropriate meeting space and for A/V equipment to play the video "Why can't you put down that phone?"
- Once you know how many people will be attending the discussion, make copies of the discussion booklet.
- Consider a hospitality table with simple refreshments, which may help to break the ice and foster discussion within the group.

- Arrive early to the session to set up the registration table and hospitality area; arrange the gathering space for the video presentation and discussion.
- Consider praying for God's blessing upon all those who will participate in your discussions.
- Consider choosing a prayer to begin each discussion session.
- The most important preparation for you as a facilitator is to go through the questions yourself. Your personal participation in the study will help you relate to and understand the experiences of your group members.
- Remember, it is not the facilitator's job to answer every question that comes up. Whenever possible, ask the participant what he or she thinks first. If appropriate, engage the others present for their thoughts.
- Stay on time and on topic. Some persons find it helpful to have a second person in the group who is responsible for giving a ten minute "warning" signal to allow ample time to finish the discussion and perhaps close with a prayer. Finishing the discussion period promptly is essential and shows respect for the participants.
- The group will expect you to keep the discussion moving.

    - Watch for clues that a timid person has something to say and encourage that person without putting him or her on the spot.

  - Kindly "rein in" participants who dominate the group. If they continue, ask them privately to help you get others to participate.
  - Gently redirect tangential remarks or questions.

- Don't share confidences outside the group.
- Enjoy yourself!

How to structure each discussion session

Here is a simple agenda:

- Introduction (5–10 minutes) for gathering and opening remarks
- View video "Why Can't You Put Down that Phone?" (5 ½ minutes)
- Group discussion (30 minutes) engaging participants using the booklet with its discussion questions, tips and suggestions and resources
- Closing remarks and possible prayer (5–10 minutes)

Section 1

Video Review Questions

1. Did you hear anything that surprised you in the video?

2. What does Marshall McLuhan mean when he says: "It's the medium more than the content that is the message"?
3. Why is *unpredictability* so powerful in stimulating desire and craving?
4. How does *operant conditioning* or *cueing* make you want to check your cell phone?
5. Persons who excessively engage in online gaming, for example, *World of Warcraft* or *Minecraft*, can have trouble stopping. Describe why this happens.
6. What is your response to hearing that: Steve Jobs, founder of the iPhone, never permitted his kids to use an iPhone and that Evan Williams, founder of Blogger, Twitter and Medium, refused to give his two young sons an iPad and instead bought them hundreds of books?[118]

Group Discussion Questions

The following questions are intended to help you reflect on your use of screen media.

1. In what ways do social media platforms and texting benefit you, your family, and friends?
2. What is your reaction when you pick up your phone to see if you have any messages or "likes"

[118] N. Bilton, Steve Jobs was a low-tech parent. New York Times. (September 10, 2014.) (Accessed on May 31, 2019.) https://www.nytimes.com/2014/09/11/fashion/steve-jobs-apple-was-a-low-tech-parent.html

and there aren't any? How does this make you feel?

3. Do you catch yourself accessing social media when you should be doing something else?
4. How much time do you spend with friends online as compared with friends in real life?
5. In what situations do you interrupt a conversation you are having with someone face-to-face to check a message you received?
6. Do you multi-task when using your screens? What do you choose to do when multi-tasking?
7. Have you ever experienced FOMO? In what situations?
8. How much personal time do you daily spend on your electronic device(s)?
9. Are there triggers for when you use your electronic device? When bored? When alone? When you first awake? When waiting in line? Others?
10. How does your use of electronic devices affect your mood?
11. When you use your electronic device, how often do you have a clear purpose in mind?
12. Do you take your phone to bed? How has this affected the amount of sleep you get at night?
13. What times or situations would you think it is inappropriate to engage in social media?
14. What would concern you if you were asked to turn off your devices for a day?
15. What skills or talents could you develop if you were spending an hour less on social media each day?

Section 2

Be a Smarter Screen User

We know the tricks that advertisers use to tempt us—a mouthwatering sundae or a shiny red car. Electronic design engineers also know how to get our attention. The color red in notifications on our cell phones and tablets triggers a desire to see what it is all about. Auto-play in videos can get us hooked to watching much longer than we intended. More and more cell phone and tablet functions are designed to get us to spend more time using these devices.

Here are a few suggestions to help you be more aware of your tech usage:

Say no to notifications. A notification is intended to let you know something important needs attention. Most phone and tablet notifications are machine-automated and do not involve people. They are intended to get you to engage with an app you might not have otherwise thought about. Set notifications only for things that are important to you.

Put your phone on "do not disturb" mode and allow only messages from "favorites." You might be surprised how much time interruptions from messages take away from what is important to you.

No screen zones. Use good screen etiquette. No screens at mealtime will help facilitate better face-to-face conversations. Also keep your phone and tablet out of your bedroom at night. You will probably get more minutes of sleep and better-quality sleep. Of course, do not read or send messages while driving.

Use an old-fashioned alarm clock. Not using your phone or tablet as an alarm clock will prevent you from being tempted to engage in the many other functions your phone and tablet offers you at a time and place designated for rest and sleep.

Utilize Apps to Help Increase Screen-free Time
Many apps are available to help you be more mindful of your use of screens and to allow for more screen-free time. Examples include *Circle Go, Qustodio, Onward*, which are app that help block websites and apps and tracks phone or tablet usage.
There are also apps that help you to designate screen-free times during the day. Some examples include:

- *Moment Family*
- *Breakfree*

Consider Internet firewalls and filtering systems for you cell phone, tablet, and computer. There are many on the market. One such app is *Covenant Eyes*.

App blocking devices to consider:

- *Freedom* for Apple and Windows products
- *Offtime* for Android products

Challenge yourself to reduce screen time.

Initially, try to reduce your screen time by one hour per week.
You could also challenge yourself to a no-screen day. Giving up screens for a day is not intended to be a punishment. It is an opportunity for you to challenge yourself to interact with people in new ways.

- Get your friends on board.
- Choose a day that is realistic to be screen-free.
- Take some time to reflect on your screen-free day.

    - What did you notice?
    - How did your no-screen day compare with a day when using screens?
    - What do you notice when you were interacting face-to-face with others for a day?
    - What were you able to do because you were not engaged with your screens?

Based upon your experience with a no-screen day, consider finding an entire weekend to go screen-free.

Need help?

For some persons, online gaming or other Internet activities become an irresistible obsession. Help is available. Treatment options range from limited outpatient therapy to intensive inpatient programs. Check online for treatment options that fit your needs.

Appendix C

Discussion Booklet for Teens

https://youtu.be/Thkl8-q2pYc

Welcome to the CMA University Video Vignette Group Discussion—"Why Can't You Put Down that Phone?" Over the course of your discussions you will view a brief video (approximately 5½ minutes), which will highlight some important principles about screen media use. After the video presentation, you will engage in a group discussion. You will have the opportunity to talk about how screen time influences your daily life. You'll hear about some ideas that have benefitted many other young adults. Some groups might want to schedule additional discussion sessions to share their experiences and ideas. Enjoy!

Section 1

Video Review

1. Did you hear anything that surprised you in the video?
2. How does *operant conditioning* or *cueing* make you want to check your cell phone? (Hint: Think of the seal in the video).
3. How can screen overuse alter the teen developing brain? What neurochemical is especially involved in this alteration?
4. Persons who excessively engage in *World of Warcraft* or *Minecraft* for example, can have trouble stopping. Describe how this happens.
5. What is your response to hearing that: Steve Jobs, founder of the iPhone, never permitted his kids to use an iPhone; and that Evan Williams, founder of Blogger, Twitter and Medium, refused to give his two young sons an iPad and instead bought them hundreds of books?[119]

Group Discussion Questions

The following questions are intended to help you reflect on your use of screen media.

---

[119] N. Bilton, Steve Jobs was a low-tech parent. New York Times. (September 10, 2014.) (Accessed on May 31, 2019.) https://www.nytimes.com/2014/09/11/fashion/steve-jobs-apple-was-a-low-tech-parent.html

1. In what ways do social media platforms and texting benefit you, your family, and friends?
2. What is your reaction when you pick up your phone to see if you have any messages or "likes" and there aren't any? How does this make you feel?
3. Do you catch yourself accessing social media when you should be doing something else?
4. How much time do you spend with friends online as compared with friends in real life?
5. In what situations do you interrupt a conversation you are having with someone face-to-face to check a message you received?
6. Do you multi-task when using your screens? What do you choose to do when multi-tasking?
7. Have you ever experienced FOMO? In what situations?
8. How much free time do you daily spend on your electronic device(s)?
9. Are there triggers for when you use your electronic device? When bored? When alone? When you first awake? Others?
10. How does electronic media affect your mood?
11. When you use your electronic device, how often do you have a clear purpose in mind?
12. How often have you gotten less sleep at night because you used your phone in bed?
13. In what situations do you think it is inappropriate to engage in social media?
14. How many of your posts help to build people up?

15. What would concern you if you were asked to turn off your devices for 24 hours?
16. What skills or talents could you develop if you were spending an hour less on social media each day?

Section 2

Be a Smarter Screen User. We know the tricks that advertisers use to tempt us—a mouthwatering sundae or a shiny red car. Tech design engineers also know how to get our attention. The color red in notifications on our cell phones and tablets trigger a desire to see what it is all about. Auto-play in videos can get us hooked to watching much longer than we intended. More and more cell phone and tablet functions are designed to get us to spend more time using these devices.

Here are a few suggestions to help you be more aware of your tech usage:

Say no to notifications. A notification is intended to let you know something important needs attention. Most notifications are machine automated and do not involve people. They are intended to get you to engage with an app you might not have otherwise thought about. Set notifications for things that are important to you.

Put your phone on "do not disturb" mode and allow only calls from "favorites". You might be surprised how much time interruptions from messages take away from what is important to you.

No screen zones. Use good screen etiquette. No screens at mealtime will help facilitate better face-to-face conversations. Also keep your phone and tablet out of your bedroom at night. You will probably get more minutes of sleep and better-quality sleep. Of course, you should not read or send messages while driving.

Use an old-fashioned alarm clock. Not using your phone or tablet as an alarm clock will prevent you from being tempted to engage in the many other functions your phone and tablet offers you at a time and place designated for rest and sleep.

Utilize Apps to Help Increase Screen-free Time
Many apps are available to help you be more aware of the screens you access and how long you use them. Consider speaking with your parents about what is available.

Utilize Apps to Help Increase Screen-free Time
Many apps are available to help you be more mindful of use of screens and to allow for more screen-free time. One example is *Onward*, an app that helps block websites and apps and tracks phone or tablet usage.

Consider Internet firewalls and filtering systems for your cell phone, tablet, and computer. There are many on the market. Some suggestions include:

- *Circle Go*
- *Qustodio*
- *Covenant Eyes*

App blocking devices to consider:

- *Freedom* for Apple and Windows products
- *Offtime* for Android products

Challenge yourself to reduce screen time. Initially, try to reduce your screen time by one hour per week. You could also challenge yourself to a no-screen day. Giving up screens for a day is not intended to be a punishment. It is an opportunity for you to challenge yourself to interact with people in new ways.

- Get your friends on board.
- Choose a day that is realistic to be screen-free.
- Take some time to reflect on your screen-free day.
    - What did you notice?
    - How did your no-screen day compare with a day when using screens?
    - What do you notice when you were interacting face-to-face with others for a day?

   - What were you able to do because you were not engaged with your screens?

Based upon your experience with a no-screen day, consider finding an entire weekend to go screen-free.

Need help?

If you know someone who is struggling with screen overuse, encourage him or her to speak with his or her parents, a youth minister or school counselor, for example.

Group Facilitator's Guide

Tips to help you plan

- A group facilitator does not have to be an expert in neurophysiology or psychology. Rather, a facilitator makes others feel welcome and is able to draw out those who need a little extra encouragement; and reins in the more exuberant participants.
- Familiarize yourself with the video, questions, exercises and suggested resources before your meeting. Spend time considering how people may respond to them. This will help you to be prepared for some directions the discussion might take.
- Schedule a 60-minute discussion session for your group as far in advance as possible and publicize it to your target audience. Because

there is more material in the booklet than can be covered in an hour, you might select questions you want your group to focus on during the hour. Let participants know that also listed in the booklet are helpful tips and suggestions are that they can consider on their own.

- Some groups might want to meet two or three times.

    - During the first session, participants can focus on questions of interest.
    - For the second session, invite participants to share some tips or recommendations that they implemented.
    - If a third discussion session is desired, focus on areas of interest and encourage ongoing implementation of tips, recommendations or other helpful ideas that emerged during the group discussion.

- Arrange for an appropriate meeting space and for A/V equipment to play the video "Why can't you put down that phone?"
- Once you know how many people will be attending the discussion, make copies of the discussion booklet.
- Consider a hospitality table with simple refreshments, which may help to break the ice and foster discussion within the group.
- Arrive early to the session to set up the registration table and hospitality area; arrange

the gathering space for the video presentation and discussion.

- Consider praying for God's blessing upon all those who will participate in your discussions.
- Consider choosing a prayer to begin each discussion session.
- The most important preparation for you as a facilitator is to go through the questions yourself. Your personal participation in the study will help you relate to and understand the experiences of your group members.
- Remember, it is not the facilitator's job to answer every question that comes up. Whenever possible, ask the participant what he or she thinks first. If appropriate, engage the others present for their thoughts.
- Stay on time and on topic. Some persons find it helpful to have a second person in the group who is responsible for giving a ten minute "warning" signal to allow ample time to finish the discussion and perhaps close with a prayer. Finishing the discussion period promptly is essential and shows respect for the participants.
- The group will expect you to keep the discussion moving.

    - Watch for clues that a timid person has something to say and encourage that person without putting him or her on the spot.

  - Kindly "rein in" participants who dominate the group. If they continue, ask them privately to help you get others to participate.
  - Gently redirect tangential remarks or questions.

- Don't share confidences outside the group.
- Enjoy yourself!

How to structure each discussion session

Here is a simple agenda:

- Introduction (5–10 minutes) for gathering and opening remarks
- View video "Why Can't You Put Down that Phone?" (5 ½ minutes)
- Group discussion (30 minutes) engaging participants using the booklet with its discussion questions, tips and suggestions and resources
- Closing remarks and possible prayer (5–10 minutes)

Section 1

Video Review

1. Did you hear anything that surprised you in the video?

2. How does *operant conditioning* or *cueing* make you want to check your cell phone? (Hint: Think of the seal in the video).
3. How can screen overuse alter the teen developing brain? What neurochemical is especially involved in this alteration?
4. Persons who excessively engage in *World of Warcraft* or *Minecraft* for example, can have trouble stopping. Describe how this happens.
5. What is your response to hearing that: Steve Jobs, founder of the iPhone, never permitted his kids to use an iPhone; and that Evan Williams, founder of Blogger, Twitter and Medium, refused to give his two young sons an iPad and instead bought them hundreds of books?[120]

Group Discussion Questions

The following questions are intended to help you reflect on your use of screen media.

1. In what ways do social media platforms and texting benefit you, your family, and friends?
2. What is your reaction when you pick up your phone to see if you have any messages or "likes" and there aren't any? How does this make you feel?

---

[120] N. Bilton, Steve Jobs was a low-tech parent. New York Times. (September 10, 2014.) (Accessed on May 31, 2019.) https://www.nytimes.com/2014/09/11/fashion/steve-jobs-apple-was-a-low-tech-parent.html

3. Do you catch yourself accessing social media when you should be doing something else?
4. How much time do you spend with friends online as compared with friends in real life?
5. In what situations do you interrupt a conversation you are having with someone face-to-face to check a message you received?
6. Do you multi-task when using your screens? What do you choose to do when multi-tasking?
7. Have you ever experienced FOMO? In what situations?
8. How much free time do you daily spend on your electronic device(s)?
9. Are there triggers for when you use your electronic device? When bored? When alone? When you first awake? Others?
10. How does electronic media affect your mood?
11. When you use your electronic device, how often do you have a clear purpose in mind?
12. How often have you gotten less sleep at night because you used your phone in bed?
13. In what situations do you think it is inappropriate to engage in social media?
14. How many of your posts help to build people up?
15. What would concern you if you were asked to turn off your devices for 24 hours?
16. What skills or talents could you develop if you were spending an hour less on social media each day?

Section 2

Be a Smarter Screen User. We know the tricks that advertisers use to tempt us—a mouthwatering sundae or a shiny red car. Tech design engineers also know how to get our attention. The color red in notifications on our cell phones and tablets triggers a desire to see what it is all about. Auto-play in videos can get us hooked to watching much longer than we intended. More and more cell phone and tablet functions are designed to get us to spend more time using these devices.

Here are a few suggestions to help you be more aware of your tech usage:

Say no to notifications. A notification is intended to let you know something important needs attention. Most notifications are machine automated and do not involve people. They are intended to get you to engage with an app you might not have otherwise thought about. Set notifications for things that are important to you.

Put your phone on "do not disturb" mode and allow only calls from "favorites". You might be surprised how much time interruptions from messages take away from what is important to you.

No screen zones. Use good screen etiquette. No screens at mealtime will help facilitate better face-to-face conversations. Also keep your phone and tablet out of your bedroom at night. You will probably get more minutes of sleep and better-quality sleep. Of course, you should not read or send messages while driving.

Use an old-fashioned alarm clock. Not using your phone or tablet as an alarm clock will prevent you from being tempted to engage in the many other functions your phone and tablet offers you at a time and place designated for rest and sleep.

Utilize Apps to Help Increase Screen-free Time
Many apps are available to help persons be more mindful of their use of screens and to allow for more screen-free time.

Consider Internet firewalls and filtering systems for cell phones, tablets, and computers. There are many on the market. Some suggestions for parents include:

- *Circle Go*
- *Qustodio*
- *Covenant Eyes*

Challenge yourself to reduce screen time.

Initially, try to reduce your screen time by one hour per week.

You could also challenge yourself to a no-screen day. Giving up screens for a day is not intended to be a punishment. It is an opportunity for you to challenge yourself to interact with people in new ways.

- Get your friends on board.
- Choose a day that is realistic to be screen-free.
- Take some time to reflect on your screen-free day.

    - What did you notice?
    - How did your no-screen day compare with a day when using screens?
    - What do you notice when you were interacting face-to-face with others for a day?
    - What were you able to do because you were not engaged with your screens?

Based upon your experience with a no-screen day, consider finding an entire weekend to go screen-free.

Need help?

If you know someone who is struggling with screen overuse, encourage him or her to speak with his or her parents, a youth minister or school counselor, for example.

General Resources

Below are suggested resources. No endorsement of all material contained therein is intended.

Websites/Blogs/Podcasts

"Why Can't You Put Down that Phone? https://cathmed.org/media - public service videos, https://youtu.be/Thkl8-q2pYc, by Sister Marysia Weber, RSM, DO (2018) Accessed May 31, 2019.

Alliance for Childhood's Critical Look at Computers in Childhood. (http://www.allianceforchildhood.org/fools_gold) Accessed May 31, 2019.

Common Sense Media. (http://www.commonsensemedia.org) Accessed May 31, 2019.

Campaign for a Commercial-Free Childhood — Every May, schools across the country participate in

"National Screen-Free Week" (http://www.screenfree.org) Accessed May 31, 2019.

Internet Live Stats (http://www.internetlivestats.com)

Social Media Use 2018 (http://www.pewinternet.org)

Screenagers Growing Up in the Digital Age (http://www.screenagersmovie.com) Accessed May 31, 2019.

TED Talks on Screen Use:

- "Why Screens Make Us Less Happy," Adam Alter (http://www.ted.com/talks/adam_alter_why_our_screens_make_us_less_happy) Accessed May 31, 2019.
- "What You Are Missing While Being a Digital Zombie," Patrik Wincent (http://youtu.be/TAIxb42FjwE) Accessed May 31, 2019.
- "Why We Should Rethink Our Relationship with the Smartphone," Lior Frenkel (http://youtu.be/Pg065s1R6TM) Accessed May 31, 2019.
- "How Social Media Makes Us Unsocial," Allison Graham (http://youtu.be/d5GecYjy9-Q) Accessed May 31, 2019.

The Art of Manliness podcast series online at http://www.artofmanliness.com/podcast

Ascension Press (http://ascensionpress.com) Accessed May 31, 2019.

Steubenville Mid-America Conference presents Paul J. Kim on elevating social media (https://steubystl365.com/elevated-living-paul-j-kim/) Accessed May 31, 2019.

Books

How to Break Up with Your Phone, Catherine Price (2018).

Reset Your Child's Brain: A 4-Week Plan to End Meltdowns, Raise Grades and Boost Social Skills by Reversing the Effects of Electronic Screen Time, Victoria Dunckley (2015).

Glow kids: How Screen addiction is hijacking our kids-and how to break the trance by Nicholas Kardara (2016).

Irresistible: The Rise of Addictive Technology and the Business of Keeping Us Hooked, Adam Alter (2018).

iGen: Why Today's Super-Connected Kids are Growing Up Less Rebellious, More Tolerant, Less

Happy—and Completely Unprepared for Adulthood, Jean M. Twenge, PhD (2017).

The Distracted Mind: Ancient Brains in a High-Tech World, Adam Gassaley and Larry O. Rosen (2016).

The Brain That Changes Itself: Stories of Personal Triumph from the Frontiers of Brain Science, Norman Doidge (2007).

The Craving Mind: From Cigarettes to Smartphones to Love - Why We Get Hooked & How We Can Break Bad Habits, Judson Brewer (2017).

Proust and the Squid: The Story and Science of the Reading Brain, Maryanne Wolf (2007).

Articles

"Does Your Child Have a Digital Addiction?" Romeo Vitelli Ph.D., *Psychology Today*, November 23, 2017 (http://www.psychologytoday.com/us/blog/media-spotlight/201711/does-your-child-have-digital-addiction) Accessed May 31, 2019.

"Do Young Children Need Computers?" Lori Woellhaf, *The Montessori Society*, (https://www.montessorisociety.org.uk/Articles/4441930). Accessed May 31, 2019.

"Phone-addicted Teens Aren't as Happy as Those Who Play Sports and Hang out IRL, New Study

Suggests," Sarah Buhr, TechCrunch, Jan 23, 2018, (http://techcrunch.com/2018/01/23/phone-addicted-teens-arent-as-happy-as-those-who-play-sports-and-hang-out-irl-new-study-suggests). Accessed May 31, 2019.

"A Nation of Kids with Gadgets and ADHD," Margaret Rock, *Time*, July 12, 2013, (http://techland.time.com/2013/07/08/a-nation-of-kids-with-gadgets-and-adhd/). Accessed May 31, 2019.

"Your Smartphone Reduces Your Brainpower, Even If It's Just Sitting There," Robinson Meyer, *The Atlantic*, August 2, 2017 (http://www.theatlantic.com/technology/archive/2017/08/a-sitting-phone-gathers-brain-dross/535476/7) Accessed May 31, 2019.

"Billionaire Tech Mogul Bill Gates Reveals He Banned His Children from Mobile Phones until They Turned 14," Emily Retter, Mirror, April 21, 2017 (https://www.mirror.co.uk/tech/billionaire-tech-mogul-bill-gates-10265298) Accessed May 31, 2019.

"Have Smartphones Destroyed a Generation?," Jean M. Twenge, *The Atlantic*, September 2017, (http://www.theatlantic.com/magazine/archive/2017/09/has-the-smartphone-destroyed-a-generation/534198) Accessed May 31, 2019.

"What's Behind Phantom Cell Phone Buzzes?," Daniel J. Kruger, *The Conversation*, March 16, 2017, (http://www.theconversation.com/whats-behind-phantom-cell phone-buzzes-73829). Accessed May 31, 2019.

"The Binge Breaker: Tristan Harris Believes Silicon Valley Is Addicting us to Our Phones. He's Determined to Make It Stop," Bianca Bosker, *The Atlantic*, November 2016. (https://www.theatlantic.com/magazine/archive/2016/11/the-binge-breaker/501122/) Accessed May 31, 2019.

"How Technology Is Hijacking Your Mind-from a Magician and Google Design Ethicist," Tristan Harris, *Thrive Global*, May 18, 2016, (http://journal.thriveglobal.com/how-technology-hijacks-peoples-minds-from-a-magician-and-google-s-design-ethicist-56d62ef5edf3). Accessed May 31, 2019.

"Reclaiming Our (Real) Lives from Social Media," Nick Bilton, *New York Times*, July 16, 2014, (http://www.nytimes.com/2014/07/17/fashion/reclaiming-our-real-lives-from-social-media.html) Accessed May 31, 2019.

"Schools that Ban Mobile Phones See Better Academic Result," Jamie Doward, *Guardian*, May 16, 2015. (http://www.theguardian.com/education/2015/may/

16/schools-mobile-phones-academic-results) Accessed May 31, 2019.

"'What's Wrong with Education Cannot Be Fixed by Technology'-The Other Steve Jobs," Tim Carmody, *Wired*, January 17, 2012. (http://www.wired.com/2012/01/apple-education-jobs/) Accessed May 31, 2019.

"Screen Time v Play Time: What Tech Leaders Won't Let Their Own Kids Do," Amy Fleming, *The Guardian*, May 23, 2015. (http://www.theguardian.com/technology/2015/may/23/screen-time-v-play-time-what-tech-leaders-wont-let-their-own-kids-do) Accessed May 31, 2019.

"19 Text Messaging Stats that Will Blow You Away," *Teckst*. (http://teckst.com/19-text-messaging-stats-that-will-blow-your-mind). Accessed May 31, 2019.

"UH Study Links Facebook Usage to Depressive Symptoms," Mellissa Carroll, University of Houston, April 6, 2015. (http://www.uh.edu/news-events/stories/2015/April/040415FaceookStudy.php) Accessed May 31, 2019.

"Suicide Rates Rise Sharply in U.S.," Tara Parker-Pope, *New York Times*, May 2, 2013. (http://www.nytimes.com/2013/05/03/health/suicide-rate-rises-sharply-in-us.html) Accessed May 31, 2019.

"Stress in America: Coping with Change, 10th ed., Stress in America Survey," American Psychological Association, online on February 23, 2017. (https://www.apa.org/news/press/releases/stress/2017/technology-social-media.pdf). Accessed May 31, 2019.

"A New, More Rigorous Study Confirms: The More You Use Facebook, the Worse You Feel," Holly B. Shakya and Nicholas A. Christakis, *Harvard Business Review*, April 10, 2017. (https://hbr.org/2017/04/a-new-more-rigorous-study-confirms-the-more-you-use-facebook-the-worse-you-feel) Accessed May 31, 2019.

"Facebook's Emotional Consequences: Why Facebook Causes a Decrease in Mood and Why People Still Use It," Christina Sagliogou and Tobias Greitemeyer, *Computers in Human Behavior*, June 2014, 35:359-363. (https://www.researchgate.net/publication/261563413_Facebook's_emotional_consequences_Why_Facebook_causes_a_decrease_in_mood_and_why_people_still_use_it). Accessed May 31, 2019.

Medical Research

"Gaming Increases Craving to Gaming-Related Stimuli in Individuals with Internet Gaming Disorder," Dong G, et al., *Biol. Psychiatry Cogn. Neurosci. Neuroimaging*, 2017 Jul; 2(5):404-12.

"Detection of Craving for Gaming in Adolescents with Internet Gaming Disorder Using Multimodal Biosignals," Kim H., et al., *Sensors (Basel)*, 2018 Jan 1; 18(1):102-114.

"Altered Functional Connectivity of the Insula and Nucleus Accumbens in Internet Gaming Disorder: A Resting State fMRI Study," Chen, C., et al., *Eur. Addict. Res.*, 2016 Jun; 22(4):192-200.

"An Update on Brain Imaging Studies of Internet Gaming Disorder," Weinstein, A., *Frontiers in Psychiatry*, 2017 Sep 29; 8:185.

"Abnormal Gray Matter and White Matter Volume in 'Internet Gaming Addicts'," Lin, X., et al., *Addict Behav.*, 2015; 40:137-143.

"Cortical Thickness Abnormalities in Late Adolescence with Online Gaming Addiction," Yuan, K., et al., *PLoS One*, 2013 Jan; 8(1):e53055.

"Impact of Videogame Play on the Brain's Microstructural Properties: Cross-Sectional and Longitudinal Analyses," Takeuchi, H., et al., *Molecular Psychiatry*, 2016 Jan 5; 21:1781-1789.

"Brain Connectivity and Psychiatric Comorbidity in Adolescents with Internet Gaming Disorder," Han, D.H., et al., *Addiction Biology*, 2015 Dec 22; 22(3):802-812.

"Abnormal White Matter Integrity in Adolescents with Internet Addiction Disorder: A tract-Based Spatial Statistics Study," Lin, F., et al., *PLoS One*, 2012 Jan 11; 7(1):e30253.

"Disrupted Brain Functional Network in Internet Addiction Disorder: A Resting-State Functional Magnetic Resonance Imaging Study," Wee, C.Y., et al., *PLoS One*, 2014 Sep 15; 9(9):e107306.

"Time Period and Birth Cohort Differences in Depressive Symptoms in the U.S., 1982-2013," Twenge, J.M., *Social Indicators Research*, 2015 Apr; 121(2):437-454.

"The Role of Compulsive Texting in Adolescents' Academic Functioning," Lister-Landman, K.M., et al. *Psychology of Popular Media Culture*, 2017 Oct 5; 6(4):311-325.

"Problematic Smartphone Use and Relations with Negative Affect, Fear of Missing Out, and Fear of Negative and Positive Evaluation," Wolniewicz CA, et al., *Psychiatry Res.* 2018 Apr; 262:618-623.

"Children's Environmental Health in the Digital Era: Understanding Early Screen Exposure as a Preventable Risk Factor for Obesity and Sleep Disorders," Wolf C, et al., *Children (Basel),* 2018 Feb 23; 5(2).

"Sleep Quality as a Mediator of Problematic Smartphone Use and Clinical Health Symptoms," Xie, X., et al., *J. Behav. Addict.*, 2018 Jan 1; 7(2): 466-472.

"Sleep Problems and Internet Addiction Among Children and Adolescents: A Longitudinal Study," Chen, Y.L. and S.S. Gau, *J. Sleep Res.*, 2016 Aug; 25(4):458-465.

"The Association Between Social Media Use and Sleep Disturbance Among Young Adults," Levenson, J.C., et al., *Prev. Med.*, 2016 Apr; 85:36-41.

"Effects of Internet and Smartphone Addictions on Depression and Anxiety Based on Propensity Score Matching Analysis," Kim, Y.J., et al., *Int. J. Environ. Res. Public Health*, 2018 Apr 25; 15(5).

"Sleep Habits and Pattern in 1-14 Years Old Children and Relationship with Video Device Use and Evening Use and Night Child Activities," Brambilla, P., et al., *Ital. J. Pediatr.* 2017 Jan 13; 43(1):7.

"Adolescent Sexting Research: The Challenges Ahead," Van Ouytsel, J., et al., *JAMA Pediatr.* 2018 May 1; 172(5): 405-406.

"The Effect of Online Violent Video Games on Levels of Aggression," Hollingdale, J. and T. Greitemeyer, *PLoS One*, 2014 Nov 12; 9(11): e111790.

"Too Many 'Friends,' Too Few 'Likes'? Evolutionary Psychology and 'Facebook Depression'," Blease, C.R., *Review of General Psychology*, 2015; 9(1): 1-13.

"Association of Facebook Use with Compromised Well-Being: A Longitudinal Study," Shakya, H.B. and N.A. Christakis, *Am. J. Epidemiol.*, 2017 Feb 1; 185(3): 203-211.

"Craving Facebook? Behavioral Addiction to Online Social Networking and Its Association with Emotional Regulation Deficits," Hormes, J.M., et al., *Addiction*, 2014 Dec; 109(12): 2079-2088.

ABOUT THE AUTHOR

Sister Marysia Weber, RSM, DO, MA is a Religious Sister of Mercy of Alma, Michigan.

She is a physician, board certified in psychiatry with a fellowship in consultation-liaison psychiatry, who trained at the Mayo Clinic in Rochester, MN. She also holds a Master's degree in Theology from Notre Dame, South Bend, Indiana. She practiced psychiatry at her religious institute's multidisciplinary medical clinic, Sacred Heart Mercy Health Care Center in Alma, MI from 1988-2014. She became the Director of the Office of Consecrated Life for the Archdiocese of Saint Louis in 2014. She currently serves as a member of the Saint Louis Archdiocesan Review Board, the Child Safety Committee, is a facilitator for Project Rachel, is an Executive board member of the Saint Louis Guild Catholic Medical Association and the Institute for

Theological Encounter with Science and Technology. She also serves as Adjunct Clinical Instructor in the Department of Psychiatry at Washington University School of Medicine in Saint Louis, Missouri.

Sister Marysia offers workshops on a variety of topics including human attachment, boundaries and character development, depression and anxiety, dialogue and conflict resolution, as well as on social media and its effects on the brain for clergy, seminarians, women's and men's religious communities, parents, teachers and students. She is a formator within her own religious community. She presents on Internet pornography addiction—a Catholic approach to treatment to bishops, clergy, seminarians, religious communities, and laity throughout the United States and Europe She presented to the U.S. Bishops in Dallas TX in 1992 on "Pedophilia and Other Addictions". She was a member of the USCCB Ad Hoc Committee on Sexual Abuse in 1994-1995. Sister Marysia has presented to the Curia, Vatican City State on "Sexual Abuse of Minors by Clergy in North America" in 2002. She has served as a psychological expert consultant for the Secretariat of Clergy, Consecrated Life and Vocations, USCCB. Her publications include "Medical Aspects of Addiction"; "The Roman Catholic Church and the Sexual Abuse of Minors by Priests and Religious in the United States and Canada: What Have We Learned? Where Are We Going?"; "Pornography, Electronic Media and Priestly Formation"; Her publications in *Seminary Journal* include: "Significant Markers of Human Maturation Applied to the Selection and

Formation of Seminarians"; "The Discernment of a Priestly Vocation and the Expertise of Psychiatry and Psychology"; and "Internet Pornography and Priestly Formation: Medium and Content Collide With the Human Brain". Her book *The Art of Accompaniment: Practical Steps for the Seminary Formator* (En Route Books and Media, 2018) is available for purchase on amazon.com along with this one, *Screen Addiction: Why You Can't Put that Phone Down.* Her chapter "Guideposts for the Seminary Formator in Understanding and Assessing Levels of Preoccupation with Use of Internet Pornography and a Formative Process for Moving from Vice to Virtue" in Spiritual Husband-Spiritual Fathers: Priestly Formation for the 21st Century is pending publication. She has also produced a short video on "Screen Addiction," which has been made freely available for public access at http://www.cathmed.org/videos.

Made in the USA
Columbia, SC
26 September 2019